JAMES LOVELOCK FRGS

Caving

B. T. BATSFORD LIMITED
LONDON

B. T. Batsford Ltd, 4 Fitzhardinge Street, London W1
First published 1969
Set and printed by Bristol Typesetting Co Ltd
Barton Manor, St Philips, Bristol
7134 0303 9

Contents

The Illustrations

PHOTOGRAPHS*

* All the photographs were taken by the author.

TEXT DRAWINGS

ACKNOWLEDGMENT

My thanks are due to the following: Mr Wilf Gresty, the Stockport artist, for his sketches; Mr Stanley Gee of the Derbyshire Caving Club for archaeological assistance; British Ropes Ltd for permission to adapt drawings of rope work; Mr Dave Allsop of the Derbyshire Cave Rescue Organisation, the British Speleological Association and the Derbyshire Caving Association for assistance in compiling this volume.

J. L.

1 *How caving began in Britain*

In the beginning, the Earth solidified from a molten mass and its crust became layer upon layer of various kinds of rock. In some of the rocks of the Earth's surface, caves, subterranean passages, narrow shafts and immense chasms opened up. From the dawn of Man's history caves have excited his wonder and awe and created the earliest legends and myths. Caves provided men with their first primitive homes. Their first demons, dragons, gods and oracles dwelt in the dark recesses of the rocks. In the folklore of many lands, goblins, dwarfs, giants, strange monsters, trolls, fairies good and bad have all made their abode underground.

Before men developed the skills of homebuilding, they first lived like the animals in the cave entrances. They did not penetrate very far but they left evidence as to how they lived, in the cooking utensils, ornaments and other signs of their growing domesticity. They left their own bones behind as clues in the evolutionary mystery. And they also left behind paintings and drawings which represent Man's earliest artistic endeavours. I have explored the famous caves at Lascaux in the Dordogne beyond the tourist limits and marvelled at the artistry of Cro-Magnon man who painted these magnificent frescoes on walls and ceilings. I have shivered at the sight of the weird handprints in the Gargas Cave near Bagnères de Luchon in the Haute Garonne, each with parts of fingers and thumbs missing due to amputation. These handprints, some 200 or so, are believed to have had some religious significance. They also pose an interesting ethnological problem, for there are tribes in Central Africa and aborigines in Australia today which still practise this form of mutilation of the hands. Scientists are still trying to find out the link between the people whose homes were underground and the primitive tribes in continents far distant from Europe. In my neighbouring county of Derbyshire, I have taken part in digging which has uncovered the bones of an animal long extinct in Britain, a species which once roamed freely over our land before the great climatic changes drove it away across Europe seeking warmer climes, or killed it off. So, caves first attracted men and beasts as ready-made homes in which the first, stumbling steps in communal, sheltered living began.

Inevitably, men's curiosity led them further and further into the earth's caves. Professors or dilettante noblemen seeking knowledge or sensation overcame their fears of the unknown to venture into the impenetrable darkness. During the nineteenth century the scientific investigation of caves and their associated phenomena began. Today, it is remarkable to think that the first depth record ever achieved by men was in the Grotto di Trebiciano in the Carso Triestine in 1841, which was not beaten until 1925.

One of the earliest attempts in Britain to establish the depth of one of our better-known potholes occurred in the seventeenth century when Charles Cotton, the poetic friend of the great angler Izaak Walton, went along to Eldon Hole in Derbyshire, one of the Seven Wonders of the Peak District, and tried to work out how deep it was by throwing down a line. Before that in the reign of Queen Elizabeth the First, Robert Dudley, Earl of Leicester, bribed a serf with a bag of gold to be lowered down the hole in order to establish its depth. According to the historian Thomas Hobbes, the serf was lowered some 200 'ells' or somewhere around 750 feet. Eventually, the man was hauled out speechless and unconscious, his hair had changed colour and he died from shock not long afterwards. Eldon Hole was my first pothole, and after my first descent, I know exactly how that serf felt. Another eminent mind which considered Eldon Hole was Charles Leigh who, in his *Natural History* of 1700, advanced the theory which brooked no argument, that the hole had been created by the Flood.

Cotton's experiment with a line must have been conducted with a certain amount of poetic licence as he made the depth of Eldon just over half a mile—2,652 feet. The experience inspired him to write a long poem in which he wrote

> *. . . who dare*
> *Look down into the Chasme, and keep his Hair*
> *From lifting off his Hat, either has none,*
> *Or for more Modish Curls, casheers his own.*

He perpetuated some of the legends about Eldon including the one about the goose which fell down the hole and eventually emerged at Peak Cavern at Castleton with all its feathers burnt off, having presumably passed through the fires of Hell. Apart from the bit about passing through the fires of Hell, this legend may have been based on fact, because there is a school of thought among cave explorers in Derbyshire that there might well be a link between Eldon and the great show cave of Peak Cavern. The first actual descent of Eldon was made in 1777 by Mr John Lloyd FRS who got to the bottom and recorded in *Philosophical Transactions* that the depth was exactly 186 feet. He prosaically recorded that: 'The light, however, which came from above, was sufficient for the reading of any print.' He survived the bombardment of falling stones as he was brought out to give what was probably the first recorded description of the descent of a pothole.

With a few exceptions, the early investigations of caves were in pursuit of the bones of early men and animals. Excavations were limited to cave entrances; only occasionally did anyone delve any deeper than the limits of natural light. Lack of knowledge and imperfect equipment restricted the earliest explorations. But stemming from the search for bones and signs of prehistory there developed a curiosity in the form and structure, the very causes of caves and their associated shafts which are now called potholes. Geologists and paleontologists developed a mutual interest in how Nature had created its underland wonderland. They speculated on the source of rivers which suddenly disappeared underground or emerged from the foot of some cliff. It was natural for this growing interest to evolve into a science itself, one involving many other sciences which I will touch upon in this book. The need to get down the shafts, to travel long distances in a world of unending darkness, to follow the underground water passages and lakes and to counter the difficulties and dangers led inevitably to the evolution of the specialised techniques necessary for this form of exploration. Present-day mountaineering evolved in somewhat similar fashion. The first mountains were climbed for scientific or philosophical reasons. The brothers Jean-André and Guillaume-Antoine Deluc of Geneva went to the top of the Buet to experiment with air pressure at altitude and also to test the boiling point of water above sea level. De Saussure first went to Mont Blanc as a naturalist. The Abbé Murith climbed the Velan because, as a botanist, he wanted to find out how high plants grew on the mountains. It was only later that mountaineering became an end in itself. So it was with cave exploration, now known as speleology.

The father of modern speleology was the intrepid Frenchman Edouard-Alfred Martel, who pioneered many techniques which are still in use today. Almost certainly it was his visit to Britain before the turn of the century that gave the impetus to the caving movement over here. In fact speleology may have started in Britain out of sheer pique that a Frenchman should lead the way to the bottom of one of the country's biggest potholes. For many years Gaping Gill had had the same fascination for Yorkshiremen that Eldon Hole had for men in the High Peak of Derbyshire. It did not have the same wealth of folklore and legend but the prospect of descending this massive hole in the limestone enticed many to make the attempt.

Nowadays there are two well-known entrances to the Gaping Gill system. One is Bar Pot, an insignificant little opening at the foot of a small limestone cliff. This was not discovered or descended until September 1949. The other is the main shaft on the east side of Ingleborough which is one of the wonders of the shire of the broad acres.

The Fell Beck, chattering merrily over its stony bed, suddenly plunges into the mouth of a wide fissure in the limestone and hurtles down dramatically until its white spume is lost in the greater darkness. It is an impressive sight. It is even more impressive when one climbs down on a swaying ladder or plunges down crazily on a bosun's chair attached to a winch cable. Possibly the first recorded attempt to descend 'G.G.', as it is affectionately known in caving circles, was that of John Birbeck of Settle in the heart of the caving country. He tried in 1872. Realising the hazard from the powerful flow of water into the pothole he set about diverting the Fell Beck. With help, he dug a trench nearly 1,000 yards long along which he diverted the waters of the stream. This done, he managed to lifeline the main shaft and succeeded in descending to a ledge about 200 feet down, a remarkable achievement in those days. He had to be pulled out.

In the summer of 1895, Martel came over from Paris on a tour of the known caves and potholes of Britain and Ireland. It was the year of the formation of the first speleological club in France where the sport-cum-science had already taken a firm hold. By then, Martel had already evolved the use of rope ladders and a system of lifelining borrowed, most likely, from the early rope technique of Alpine climbers. Edward Whymper had been one of the first to write about the art of rope management and its correct use on steep rocks and on ice. Martel was also probably the first man to recognise the value of the telephone to cave explorers. One of the problems of speleology is that of communication between members of a caving party who may be separated on the various pitches of a pothole or extended through the length of a cave. In some places it is possible to use a code of whistle signals but highly efficient field telephones and occasionally radio links are more reliable and practical. Martel used a telephone in Gaping Gill. It was a primitive affair and was subject to rather more malfunctioning than is the present-day instrument.

Martel made his historic descent on 1 August, 1895 and, as this marked what might be called the birth of speleology in Britain, it is worth recording fairly fully. The news of his attempted descent brought a sizeable crowd over the moors to the pothole entrance and they watched with interest as he tied his long lengths of ladder together, measured out what he felt was sufficient safety line and, through his English-speaking wife, gave his helpers instructions on the management of the rope. He had tried several days before but the English summer was as unpredictable as ever and thunderstorms swelled the beck to alarming proportions. By noon on 1 August, however, conditions had improved sufficiently for him to make the attempt.

There was no shortage of willing hands to hold the lifeline, in fact there were so many as to represent a positive danger. With the air of a man who knew exactly what he was doing, Martel raised his cap, adjusted his lantern and stepped over the edge on to his ladder. As he made his slow descent, Mme Martel relayed his instructions to the lifeline party as they came crackling over the wires. Like a true explorer, Martel was duly cautious. He left little to chance and he paused frequently to evaluate his position. Daylight gradually faded as his practised feet sought out the rungs beneath him. He was immensely strong and possessed extraordinary stamina. This he needed when he first climbed into the waterfall. Although the Fell Beck had been diverted somewhat, it was not possible then, as it is today, wholly to prevent the stream from cascading down the shaft. There was a considerable body of water thundering down what is described as England's highest single-drop waterfall, as Martel was hit by its full force. Although anticipating it, his breath was taken away for a few minutes. I know only too well how he felt, as I have been stuck in the same waterfall myself. Even with present-day waterproof equipment, it is always an unpleasant experience. How long one stays in the wet depends on the sense of humour of the man driving the winch.

Around 190 feet from the surface, Martel touched down on the same ledge that Birbeck found twenty-three years previously. Here, he stopped while he carefully sorted out the coils of rope ladder and safety line which had piled up on the ledge. Although it looks and sounds quite easy, it is often difficult to throw a ladder correctly down the full length of a deep pitch of a pothole. It can snag and snarl on all kinds of projections on land or ledges far from the bottom. Leaning out from the ledge with the lifeline taking the strain, Martel was able to guide the ladder down the remainder of the shaft. But his light was feeble and the daylight from above was too poor to show whether or not it had actually reached the bottom. He had taken careful soundings previously and was satisfied that he had sufficient ladder for his task. But he was still uncertain. He paused to look up at the waterfall spurting just beyond him. In some lighting conditions it gives off a lovely rainbow effect, but overhead it was a dull day and his lamp, by modern standards, was a bare glimmer. The fall was deafening and ominous but Martel stepped boldly from his ledge having assured his helpers, over the telephone, that all was well and he seemed certain of reaching the bottom safely. As he descended further, the ladder began to sway alarmingly just below where the roof of the main cave abuts on to the wall of the main shaft. This motion carried Martel in and out of the waterfall, ducking him repeatedly.

The force of the water knocked him breathless. To add to his difficulties, Martel realised he had miscalculated the length of his safety line and he reached the end of it before getting to the bottom of the cave. Over the telephone, he called to his helpers to tie on another length. The roar of the water rushing past him made it impossible to tell whether his message had been heard and understood. He dangled, drenched to the skin, clinging in the icy blackness to the rope ladder. As the rope round his waist slackened, he sensed that a new length of line had been added and he set off down the last few feet to the cave floor. To the watchers above, and no doubt to Martel himself, it seemed like an age between his leaving the edge of the pothole and reporting his safe arrival at the bottom. It took, in fact, twenty-three minutes.

Calmly, over the telephone he announced: 'I have reached the bottom. I will untie myself and go and explore the cave.' Before he did so, with true Gallic appreciation of the needs of the inner man, he told his helpers, 'You can now have lunch'. While the surface party were eating their sandwiches and wondering if they would ever see Martel again, the Frenchman was placidly perambulating round the Main Chamber, which covers over half an acre and is more than 100 feet high. It must have been an awe-inspiring sight for him to look upwards. The waterfall thudded on to a pile of rocks, some of which had fallen from the shaft sides and others which had been lobbed in by passers-by. From top to bottom, the shaft is 350 feet deep and the waterfall is one of the most dramatic sights in British caving. By the faint light of his lantern, Martel probed around the Main Chamber, which is 500 feet long, noting the various passages leading away from this ante-room of the underworld. He surmised, quite rightly, that this was only the beginning of a big cave system. As later explorations proved, it is one of the biggest systems in Britain, extending for nearly four miles with at least one other entrance in addition to Bar Pot, the extremely severe Car Pot with its immensely tough, four-inch-high squirm into the East Passage of Gaping Gill.

Writing later, Martel commented: 'It was one of the most amazing subterranean scenes I have ever witnessed.' As he had already been in some of the greatest gulfs in France, this was tribute indeed to the splendours of the Yorkshire underworld. In spite of his limited equipment and poor lighting, Martel was able to make a quick preliminary survey and a reasonably accurate sketch of what he had seen. This done, his chattering teeth warned him that cold and fatigue, among the worst enemies of cavers, were setting in. Although it was summer, the cave was far from warm—no caves ever are.

He returned to his telephone, which he had tucked under a rock for safety. Eagerly he called the surface. '*Allô! Allô! Allô!*' he shouted into the instrument. '*Tirez doucement . . . Entendez-vous?*' There was no answer from those above. The line was dead. Water must have made its insidious way into the primitive handset and shorted the connections. Or the line may have been snicked by a rock. Whatever the reason, he was out of vocal contact with his helpers. But he had also arranged a few signals which were to be given on the safety line in the event of the telephone breaking down. Wisely, he tied himself on first, before giving the first tugs to let those above know he was ready to come out. I say 'wisely' because potholers who had failed to tie on before giving rope signals have found the loose end whisked out of their hands and it is not always easy to get the rope down to exactly the right spot again. No sooner had his signals been understood than the belay party started hauling away with immense enthusiasm and vigour. So much so that poor Martel's feet hardly touched the rung and he was jerked skywards like a pantomime fairy in an aerial ballet. Just as suddenly the pulling stopped. The rope had probably jammed in a crevice or become wedged between the ladder and the rock. It was certainly not a healthy place for that to happen and Martel was left to admire the scenery while the belay party tugged and tugged to free the line. To help them, Martel climbed a few rungs to give the rope some slack and this was sufficient to unwedge the rope and allow the pulling to continue unhampered. He paused for a rest on the ledge and then made the rest of the ascent without incident. At exactly 3.55 p.m. he was back at the entrance to the shaft and the muggy warmth which greets potholers on their return from the cold realms below. He bowed modestly to the cheering crowd and politely avoided the pats on the back which might have knocked him down the shaft again at a faster rate than he had recently descended it. It was an emotional occasion. His wife kissed him passionately. A hot rum was quickly provided. Dry clothes were donned. Martel was once again the immaculate Frenchman. And courteous as always. He never forgot the assistance given to him by many Yorkshire folk, whose chagrin at seeing their greatest chasm fall to a foreigner he must have sensed. He was truly grateful and, when writing of the conquest of Gaping Gill, he commented: 'My Gaping Gill public was the calmest and most sensible I ever came across.'

Martel's feat echoed throughout Britain. It was a gauntlet thrown down to Yorkshiremen on their native heath and the challenge was not slow in being accepted. The doughty Yorkshire Ramblers' Club, formed in that golden age when men were first turning to their native hills for mental and physical recreation and escape from industrial squalor, tackled the hole a few years later. They used a winch instead of a ladder and started the series of explorations which have now uncovered most of the secrets of this truly remarkable cave.

Perhaps it was not without significance that the weather should shed a few tears on Martel after he made his descent. Only twenty minutes after his return to the surface a violent storm broke over Ingleborough swelling the beck so much that, had Martel been on his ladder then, he could easily have been swept away. Today, Gaping Gill is classified as a Super Severe Pothole in the graduated standards of difficulty by which caves and potholes are known. I have been down there for five days on a big exploration. My friend Geoffrey Workman established the first British stay-down record in Gaping Gill when he remained alone, camping in the Sand Cavern and carrying out a great deal of research into the cave and also into human behaviour in isolation underground.

Speleology in Britain began with Martel. Ironically, it was a British team which beat the French at their own game when they achieved the caving equivalent of climbing Everest, establishing a new world depth record in the Gouffre Berger, the deepest known cave in France, which was first descended by the French.

2 *How to cave: personal equipment*

In the days before cave exploration became as highly organised and scientific as it is now there was a fairly light-hearted and casual approach to sojourns underground. A few old clothes, nailed boots, a torch or a few candles and a length of hemp rope—and the inevitable flat cap—were adequate enough items of equipment for the pioneers of caving. With these unsophisticated items the early speleologists achieved wonders. But as with mountaineering and rock-climbing generally, great strides have been made in the evolution of new equipment and techniques designed to give a greater measure of safety as well as extending the limit of what is possible.

Personal preference is involved in the selection of one's equipment but there are certain basic requirements when it comes to choosing gear for caving. As other people's lives and limbs may be jeopardised as well as your own if you are ill-equipped, it is as well to acquire the best you can afford—and learn how to use and maintain it. Starting from the top, the most obvious essential is a helmet to protect the skull from the inevitable bumps and bangs to the head while underground. A helmet also provides protection against rock falls and is the best platform for lights, leaving both hands free for climbing, crawling, rope and ladder work. With the development of modern plastic materials, protective headwear has improved enormously and it is both light and strong. Shops now exist which cater for the caver—if not exclusively, adequately enough to provide all his needs. From these, caving helmets are readily available. A caving helmet, with care, should last a lifetime. The best sold by reputable firms come up to the British Standards Institute specifications and are light, yet strong enough to withstand the most rigorous treatment. These are marked with the BSI 'Kite' mark as proof of quality. I can particularly recommend the Texolex, No 2 National Coal Board helmet which has been designed with speleologists in mind. It has the appropriate lamp bracket and air holes which help to keep the head cool. Some people prefer helmets without vents, using them for baling and other bizarre functions which have nothing to do with the protection of the head, but I prefer a helmet which 'breathes'. Take care over the fit of the head-band. I think it is better to have a fairly tight fit than one which is loose. An ill-fitting helmet is apt to flop about and drop over the eyes as well as making disconcerting shifts of the light beam. It can also fall off and this could be disastrous. A too-lengthy peak can also restrict vision but, if it is too long, the neb can be shortened with a saw.

As the caving scene is one of unending darkness, except in show caves where electricity is usually laid on, portable, artificial lighting must be carried. The determining factors are the usefulness of the equipment for the task in hand. The earliest explorers used candles and these are still a useful and dependable part of any caver's auxiliary or survival lighting equipment. The best are the chubby, long-life type which last up to seven hours. Next comes the good old carbide lamp which operates by burning acetylene gas. An extremely efficient form of carbide lamp is easily purchasable. It has a fling-wheel ignition, a good reflector and a hook to hang it in position on the helmet. They come with small and large reflectors. Some cavers prefer carbide lamps to electric lighting because, once they become familiar, they give a fairly large area of light, equally diffused, which is better to work by than some electric lamps that give a powerful beam; although it may show a long way ahead, it tends to give inadequate lighting where it is often needed, close to the hands and feet. Carbide lighting is less hurtful to the eyes of others when directed towards them and the glow from an acetylene lamp is softer and warm-looking. The flame, which is adjustable to a certain extent, can be used for warming the hands, heating soup or detecting the presence of heavy concentrations of carbon-dioxide which is present in some lead mines. But acetylene lamps have a number of disadvantages. Water will, of course, put them out, so they cannot be used for diving under ducks. They can also be extinguished by strong draughts. Care must also be exercised in the use of carbide lamps close to such man-made fibres as nylon, Ukstron and Courlene which are widely used in caving. The naked flame can melt and weaken these ropes. In confined spaces, acetylene lamps tend to burn up oxygen fairly rapidly, which can affect the breathing. And the gas itself, if allowed to escape in a confined space, could have unpleasant results. Acetylene itself may not be lethal but its presence can cut down the oxygen supply to below that of the body's minimum requirements. It is also more than likely that it will contain traces of phosphine which is, of course, highly dangerous and extremely toxic. I have no doubt that the carbide light he had on his helmet was one of the contributory factors in the death of Neil Moss in Peak Cavern, Castleton, Derbyshire, in March 1959. It drastically reduced his oxygen supply and contributed to the anoxia which led ultimately to his tragic end and entombment for ever in the Derbyshire limestone.

As carbide needs water to activate it into a gas, water must be carried or be available underground to replenish the lamp's reservoir. The carbide itself should not be too large and the best is that which is up to British Standards specification 642. Cleanliness is the most important part of maintaining a carbide lamp. It must be thoroughly cleaned after use, and the flint should also be checked and changed if necessary. And never leave your spent carbide in a cave and pick up any new carbide you may spill when recharging the base of the lamp, as it can poison all forms of cave life.

There is a wide variety of electric lighting suitable for caving, both wet- and dry-cell. Highly popular these days is the accumulator type of lighting set used by miners. It is the most expensive form of lighting and requires a battery charger to revitalise it. These sets may be bought new but there are also on the market reconditioned National Coal Board accumulators which cost about £2. The head light contains two bulbs—one which gives full power for about ten hours and the other which uses half power and lasts for nearly twice as long. This type of lamp is useful in so far as the half power can be used for a general working light and the full beam for looking down deep shafts or long passages. The main disadvantage of this equipment is its weight—somewhere around eight pounds—and it is also bulky. Worn normally on the hip, the accumulator batteries tend to catch on various projections. This equipment cannot, of course, be recharged underground unless one has a portable generator, which is unusual underground. This type of lighting needs a good deal of maintenance. It must be regularly charged, topped up and checked for leakages. Bulb replacement requires a special key and this must be included among the normal spares when going underground in case a bulb needs changing.

In addition there are a number of dry-cell battery lamps which include a head-band for fixing to the helmet. They use a varying number of U2-type batteries and care must be taken to insert them in the battery case in the right order of polarity. The battery carrier must have a belt clip. On the ones usually sold, the clip is really too weak for the strain of caving and needs reinforcing. When changing batteries underground, always take away the spent ones as these, like carbide, apart from looking unsightly, can endanger the sensitive forms of cave life.

Another useful battery is the doorbell type, which has two brass screw terminals and is of a convenient size to fit into a breast pocket. I have used this type for many years and have found it quite satisfactory. As it tends to break up when the cardboard container gets wet, it it a good idea to protect it and bind the cells more firmly together by using insulation tape to completely cover the battery. Polythene sheeting, made into a form of pouch, can also be used to protect this type of battery.

Torches are handy as extra lighting for poking into crevices, etc., but I do not recommend their use as a main lighting as it is important to have both hands free at all times for ladder work, crawling and handling equipment. The heavy moulded rubber type is best for caving as it is best able to withstand hard knocks and submersion.

For all these forms of lighting I have mentioned, both carbide and electric, it is vital to carry spares. Carbide is needed for replenishing the acetylene lamps and water must be carried too, if none is available underground. For long trips, a pricker should be included to clean the jet. Electric lights need spare batteries and bulbs. The bulbs should be wrapped in cotton wool and can be carried in the usefully sized 35-mm. film cassette tins. These tins can also be used for carrying matches and it is worthwhile popping in a little bag of silica gel to help absorb any moisture there may be in the air. And do not forget to take a small piece of sandpaper in each match-carrying tin as there is nothing worse than having matches and no means of igniting them. The rocks underground are frequently damp and cannot always be used for striking purposes. Providing it is dry, your thumb nail, snatched across the match head, can also be used. And a cigarette lighter is another useful accessory.

I learnt very early in my caving career that the right kind of clothing makes caving easier, safer and more comfortable. On my first caving trips I used an old shirt, sweater, jeans and a balaclava helmet. In the crawl at the bottom of Eldon Hole I wriggled out of my jeans and one of the most complicated manoeuvres I have ever made was reversing into them again. Caving clothing must fulfil two main functions—to preserve body heat and prevent exposure. Cavers will never win any prizes for sartorial elegance. I have seen some remarkably bizarre costumes underground. There is plenty of scope for individual eccentricity and this is, perhaps, the one human circumstance not catered for by Moss Bros. Some people swear by wool next to the skin—and

1 Basic caving equipment

2 Karabiners or snap rings with screw fastenings of the gate, best for caving

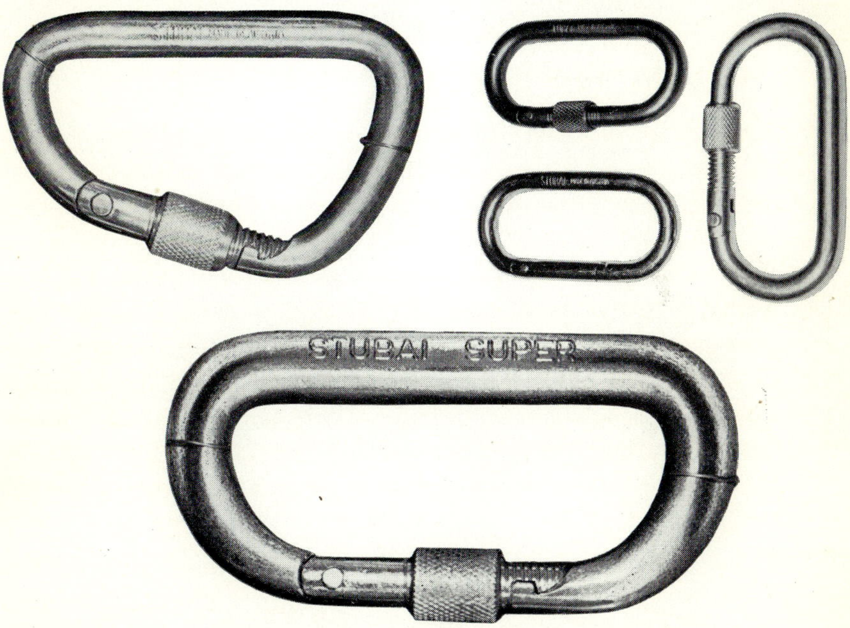

3 A well-equipped caver—overalls, wet suit, gloves for protection on wire ladders, waist sling and karabiners

4 A bag with a draw cord is best for transporting potholing equipment underground

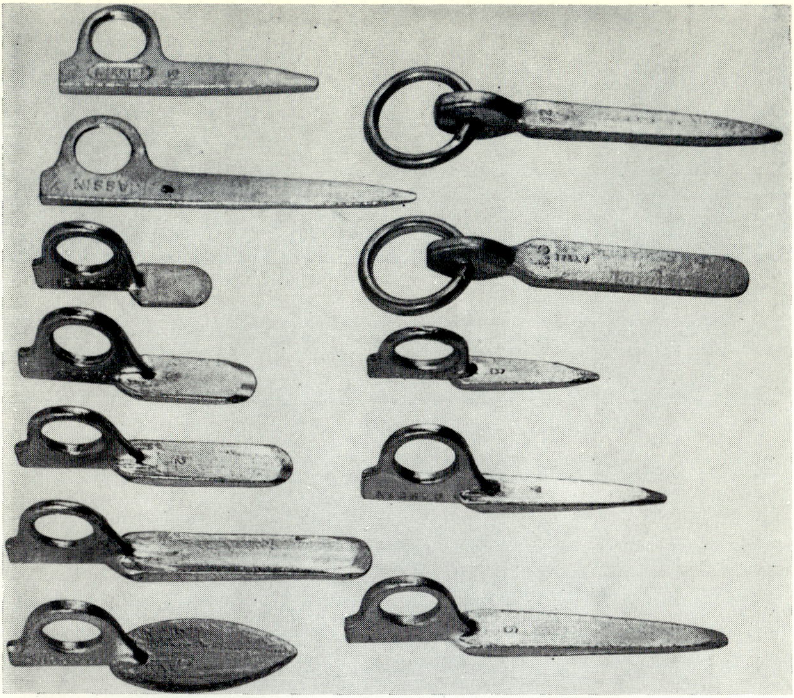

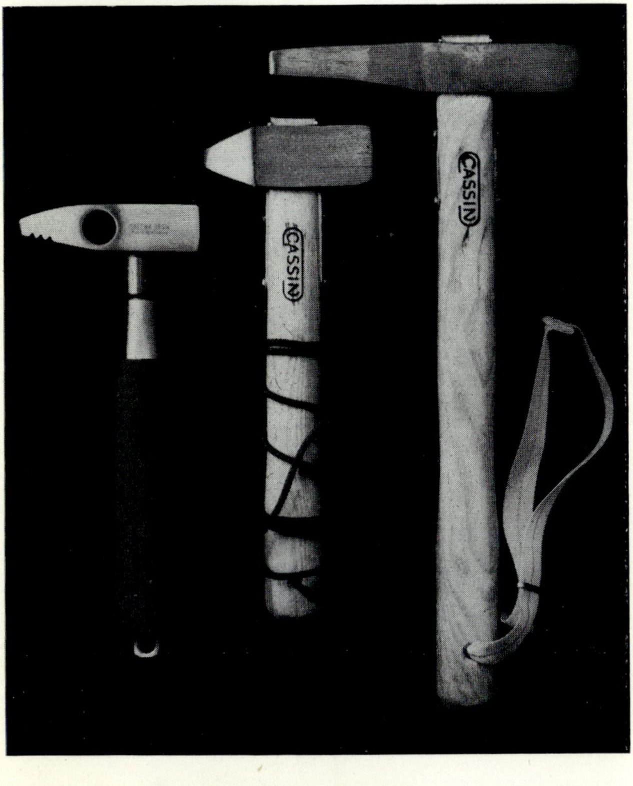

5 (*above*) Various pitons made from mild steel suitable for caving purposes. 6 (*left*) Typical piton hammers useful for caving

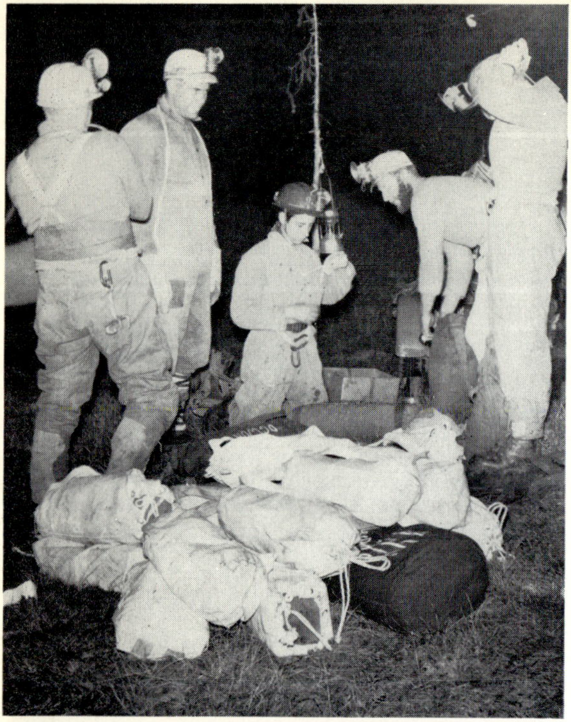

7 *(left)* A typical expedition scene as a caving party prepares to make a night descent of the Gouffre Berger, France. 8 *(below)* Cavers pumping out the unexplored entrance to a cave in Derbyshire

some people cannot stand it. I once had a long talk with a doctor about clothing and exposure. 'Keep the viscera warm, lad. Keep the viscera warm,' he said patting my abdomen firmly. I have found this extremely sound advice. Consequently a pair of woollen bathing trunks is a good way of keeping the lower abdomen warm. Some people prefer string vests to woollen ones; again this is a matter of choice as is the wearing of long string underpants or Long Johns, and one old caver I know would not be without his good old-fashioned combinations as a useful, one-piece, all-covering undergarment. Flannel shirts are preferable to cotton or nylon and two or three light woollen sweaters are better than one big, chunky one. The sweaters can be added or discarded as needed. Try and find the happy medium which is compatible with warmth and mobility. It will be a case of trial and error before you strike the ideal balance. Over-bulky clothing makes movement difficult and increases the risk of jamming in narrow passages. And whatever one wears underneath, a one-piece boiler suit or overall is a 'must'. This must be big enough to cover a reasonable amount of clothing and have this in mind when buying one. There should be freedom to move the hands above the head and stretch the legs and bend the knees easily. Nylon overalls, specially made for the purpose, have been tried with success as they do not hold water to the same extent as cotton and dry much more quickly on the body or when hung up. If you are having one made, one pocket on the chest is the only requirement. The average boiler suit has many pockets which are best left empty as things in them bunch up and restrict movement. A leather belt is useful to hang a battery clip on and is needed if chest harness is chosen for protection while climbing. Two pairs of woollen socks, preferably knee length, should be worn, one pair under the overall and the other pulled up to the knee on the outside. Some cavers like to wear anklets or gaiters which are made of canvas or nylon with instep fastenings. These cover the tops of the boots and the calves and help to prevent trouser legs from ruckling and snagging on ladders, rocks and other projections. These, again, can be fairly expensive but can be made with a little ingenuity by a willing wife or girl-friend. Alternatively, you can get someone to knit you woollen tubes or cut the feet off old, long socks—these will suffice to tuck the trouser legs into. But beware of having anything too tight round the legs as this can cause restrictions to the circulation which can be harmful in many ways.

In caving, one must become resigned to getting wet. It is virtually impossible to keep dry. Some cavers try to postpone the evil moment as long as possible and go to elaborate lengths to keep out of water. Some will only go in dry caves for that reason. But water is one of the fascinations of the underground and cannot really be avoided by the serious explorer. But the chilling effect of water must be guarded against in order to keep exposure at bay. There are a number of ways of doing this. One is to wear one of the former RAF exposure suits, of which there is still a plentiful supply in this country. Known also as 'Goon Suits', they were never designed for caving, being of thin rubber, but they do give some protection and comfort. It is best to cut the feet off as they are big and floppy and were not intended for walking. A band of inner-tubing will help to keep a lot of water out of the legs. Most of these suits are inflatable but there is little value in blowing them up unless one wants to float down a long, underground river. I did this once, with an expedition in France, and the party looked like so many Michelin men bobbing along like corks in the current. But these suits puncture easily and, in some cases, have been known to cause exhaustion by overheating the body during extreme exertion. There is also the danger of shock should water suddenly enter when the body is perspiring. Nevertheless, Goon Suits are fairly cheap and afford reasonable protection. They do have cords and other things dangling from them and it is best to cut these off and wear the whole thing under an overall.

The other most useful form of protection against water is the wet suit made from Neoprene foam rubber. This is a close-fitting sheath which fits snugly to the body. Prices for these range from just over £6 to around £15 and they can be bought or made up to your own size in do-it-yourself kits. There are three gauges of varying strength and thickness and they do away with the necessity for wearing much bulky underclothing. A popular make of this suit is the Castleton kit made by Sirocco Suits of Scarborough, who have specially designed these suits for cavers, coming, as they do, in five sizes plus tapes, adhesive (useful for repairs) and zip fastening. On the face of it wet suits are dear, but if you do intend to take up caving seriously and spend a lot of time in water passages, climbing up and down waterfalls and diving under ducks, the wet suit is an investment.

Another important item of equipment is boots. Again, no ideal caving boot has yet been evolved. But boots should be of good-quality leather to withstand the constant soakings and, if you intend them to last, it is worthwhile cleaning them after every trip, washing off the mud and soaking them regularly with dubbin or Mars oil. The Commando- or Vibram-type sole is best, although it slips on wet and muddy limestone. Some cavers put in the odd tricouni plate of a couple of clinkers round the edge of the boot as a better aid to climbing rock pitches, but these are not essential, as ladders are more often used for climbing. If you dislike having water in your boots, you can drill a few small holes in the sole to let the water out. Boots with steel toe-caps best stand up to the strain of crawling and the battering they get catching against rock walls when on ladders and are widely used. Avoid boots with clip-over hooks for lacing as the hooks catch on the thin wire of the metal ladders now in use and can cause serious upsets. I am not entirely happy either about boots with D-rings for the lacing system, as these too can catch on the wire and rung ends. The old eye-hole type are undoubtedly the best and give less trouble. Nylon laces are best as leather becomes brittle and snaps.

What I call optional extras are gloves. Industrial gloves are of use in protecting the hands when hauling loads or lowering equipment. They can be used on wire ladders where extensive climbing can cause nasty cuts and abrasions to people with soft hands. For ladder work, the gloves must be close-fitting and supple. There are also special belay gloves now available. They are of leather and are fingerless and have been designed to give a positive grip on a rope under wet or dry conditions. And they eliminate friction burns while lowering equipment or in the event of a fall.

For people with tender knee caps (haven't we all) miner's knee pads of leather and plastic with leather binding straps are good things to wear to take the agony out of 'knee-wrecker' crawls. They may be condemned as 'cissy' but unless one is an absolute masochist, I think it is a good psychological factor to eliminate as much pain and discomfort as is humanly possible.

Last, but not least, come articles of personal equipment for securing oneself to ladders, rocks, lifelines and belay points. When climbing on rock or on ladders, it is the first rule of safety to be secured by a lifeline to guard against a fall. I will go into belay techniques elsewhere but one's own personal gear should include a waist-band or the baudrier or shoulder harness. My fellow climbing club member Ken Tarbuck recommends a waist-band of three-quarter inch Italian hemp about twenty feet long, wound several times round the body. This allows the free movement of the karabiner or snap ring and is less harmful to the body than a single loop. It needs tying with the waist-band tie (*see* p. 46) with a reef knot with the overlapping ends tucked into the lay of the rope. This rope is a useful extra length when needed. Alternatively, the waist-band can be of nylon five-eighths of an inch thick. Also becoming popular these days is the shoulder harness or baudrier which is a loop of rope fastened diagonally across the chest and secured with a sheet bend into which the karabiner is clipped. Also on the market are bandoleers or chest harnesses specially designed for the job. They have been evolved for artificial climbing but can be usefully applied to caving as they lessen the risk of internal injury and spread the load, in the event of a fall, over the rib cage. These chest harnesses are much better when it comes to affixing oneself to a ladder for a rest or in an emergency. From a caving point of view, their disadvantage is that they add a little extra bulk to the wider part of the body, but they are worth using for their greater security.

I have already mentioned karabiners or 'krabs' as they are more popularly known. These can be either of steel or alloy and have a spring-loaded gate which can be opened and closed to allow a rope to be attached. These come in various shapes and sizes. The strongest and most secure are the D type, with a screw which covers the gate, making it impossible for the rope to jump out, and prevents the accidental opening of the gate. A good British model is the Hiatt D-shaped made of steel with a screw-gate, which weighs just under five ounces and has an approximate breaking load of 5,000 pounds. This costs around 13s 6d and compares with the German Asmu chrome vanadium steel which has a breaking load of 7,500 pounds and costs about twice as much. Cassin and Stubai make most of the lighter alloy karabiners which have a slightly reduced breaking load. Again, a good karabiner is an investment in life insurance. Do not use, under any circumstances, any of the cheap ex-War Department surplus karabiners which can be bought at some surplus stores. These are notoriously unsafe, being weak and prone to collapse, and have been condemned by mountaineering authorities. Always clean your 'krab' and oil the gate action frequently. Nowadays, many cavers make rapid descents by ignoring the ladder and rope down the various pitches. It is a technique known as abseiling or rappelling. One simply slides down the rope which, in the classic method, involves the use of a double rope looped over the body and thigh which absorbs some of the friction with the hands controlling the speed of the descent. There is the karabiner method in which a sling is used and the karabiner fastened to it in the crutch and the abseil rope passed through the karabiner, over the shoulder and across the back. And the third method brings me to another personal gadget which is extremely useful on long descents, the trident or descendeur, which is a three-pronged

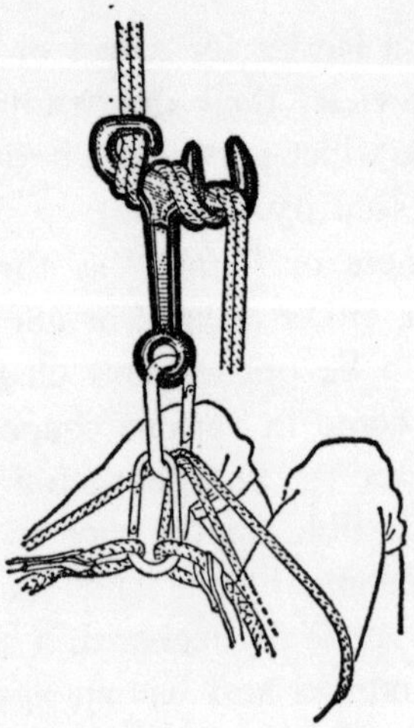

1

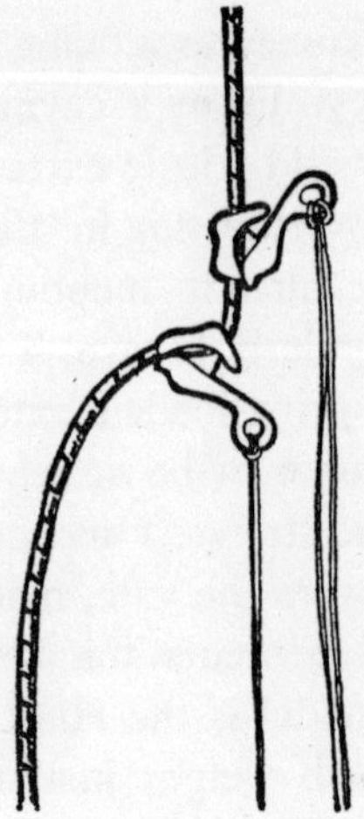

2

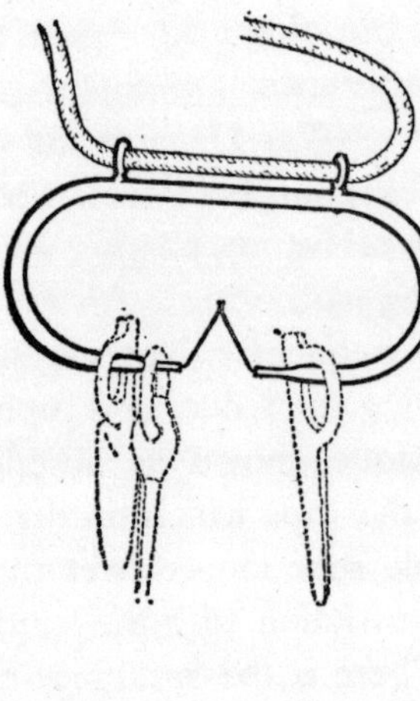

3

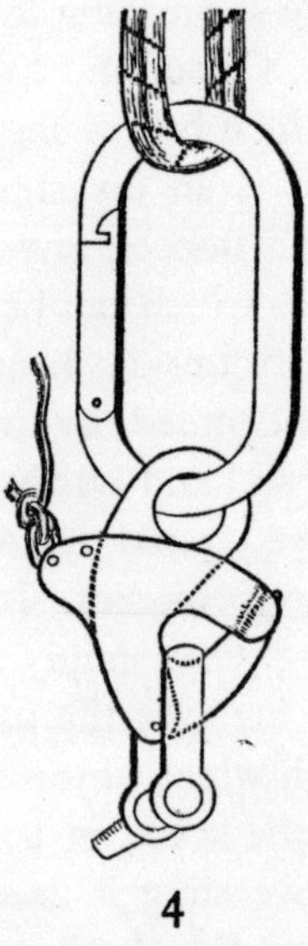

4

1 The Pierre Allain descendeur used for quick, double rope descents. Useful on long abseils.
2 The Salewa prusikers which take some of the strain out of prusiking.
3 Piton carrier with spring clip for quick release.
4 Universal fifi with safety latch which can be pulled free from any angle.

fork which clips to the waist-band. The abseil rope wraps round the descendeur and is again controlled by hand.

Descendeurs come into their own where many pitches are involved and save a lot of wear and tear on clothing.

Other items of personal equipment should include a whistle with which to give signals for being raised or lowered on ladders. This should be fastened on a neck-band and carried inside one's boiler suit and never loosely outside as this can lead to possible strangulation should it get hooked up somewhere. A knife is also worth carrying, preferably one with such gadgets as a tin-opener, screwdriver, splicing tool and corkscrew. All these can come in useful underground. A personal first aid kit should also be carried in addition to a larger kit for a group. This should consist of dressings for minor cuts, anti-burn ointment, headache tablets, preferably the more easily absorbed type and not aspirin which is harmful to the stomach lining, among other things. And for note-taking and surveying a notebook and pencils, chinagraph crayon and a small sheet of plastic on which to write should be taken.

All the personal gear which is not wrapped around you should not be stuffed into pockets but carried in a shoulder bag. An ex-WD canvas bag is ideal and plastic can be used to keep the articles dry inside it. Some people carry things in tins but these can be awkward and unyielding in narrow places, and the sharp edges pressing into the body can be painful.

These items, then, are what one needs personally to assemble. The other equipment I shall describe is that usually provided by a group or club.

3 *How to cave: group equipment*

I come now to what properly may be called group or club equipment. This is usually equipment beyond the financial resources of the average caver and is bought collectively by the club for general use. This should be under the care of one person, often the most important member of a club or group, the tackle master. The correct stowage and maintenance of equipment must never be a casual affair and an experienced and thoroughly responsible person should be chosen for the task.

Many cavers prefer to buy their own ropes for lifelining. I think it is a good thing to maintain one's own rope enabling a careful check to be kept on its condition and eventual renewal. But for the bigger expeditions a great deal of rope is needed. There are many arguments about the right choice of rope. As in the world of fashion, man-made fibres have made a dramatic appearance and have almost completely superseded natural fibre ropes which are affected by damp and other factors.

Nylon has come to be accepted as a good all-round caving rope and some are specially designed with caving in view, such as British Ropes' Formula S, which is light and floats on water. It is unaffected by mildew, oil, alkalis or acid and has considerable resistance to abrasion. Its wet strength is superior to many of the other polypropylene ropes. With a one-and-a-quarter inch diameter, it has a breaking strain of 3,000 pounds. Some cavers think that one-and-a-quarter inches is too narrow, especially if any heavy lifting has to be done, and that a better grip can be had on the thicker ropes. Viking No. 4 has this additional quality with a consequently higher breaking strain of 4,200 pounds. Danger to nylon occurs from internal abrasions set up in the fibres by accumulations of grit which is picked up during normal wear and tear in caves. For this reason, nylon ropes need regular inspection and they should be discarded at the first sign of deterioration. Never say 'Let's risk it'—it could be disastrous in a crisis. It is a good idea to have a small tag on each rope indicating when it was bought. A log book of ropes should be kept recording their usage and condition. If a rope has been subject to a severe strain such as arresting a fall or hoisting a heavy load, it should be checked before re-use or discarded entirely. Do not leave it lying around or it may be taken out accidentally and fail when most needed. Some clubs have adopted the good idea of fixing a time limit on their ropes and throw them out automatically every two years.

Recently ropes other than nylon have made their appearance. Ulstron has been used successfully in the Gouffre Berger by a British team, and Courlene as well. The new sheath-cover or kernmantle ropes of perlon are also finding their way into caving use. These have a graded outer sheath which protects the inner core against abrasion and other damaging influences, and are made both in England and by Continental manufacturers. But this type of rope has as yet to be fully proved. Some think that although it has a high tensile strength, it does not absorb as much energy as the other ropes, although it has passed the rigorous international Alpine Clubs' Dodero test.

None of these ropes, incidentally, should be used in the manufacture of ladders because they have too much stretch due to their elasticity. In fact rope ladders are now things of the past. Natural-fibre and wood ladders are still in use but are rarely made nowadays. They take up a lot of space when rolled up, they absorb moisture and mud and become exceedingly heavy, the rungs are more prone to breakage and they are not always easy to join and unjoin. It is sad for me to have to say this, as I first learnt my ladder work on the good old-fashioned Yorkshire ladder with satisfyingly thick rungs, widely spaced. Today electron or wire ladders have come into universal use. Although they look frail, they are extremely strong and stand up better than natural-fibre ladders to the stress of cave life. They were developed in France and gradually caught on in Britain. Basically, they are made from galvanised steel wire with a breaking strain in excess of 1,100 pounds with six-and-a-half-inch-wide alloy rungs. They are manufactured in twenty-five feet lengths (or longer if necessary) and cost about £7-£8 to buy. Many clubs prefer to make their own ladders as it is a useful group activity. But it is a highly skilled business requiring the use of a fly press, infinite patience and a knowledge of basic metallurgy. In manufacture only the finest galvanised wire should be used.

Rungs should be between ten and twelve inches apart and standardised throughout the club's ladder equipment. Ladder tails (where they are joined to other ladders) should be half the space between the rungs so that, when they are joined by the C-ring fastenings, they make a regular-sized step the same size as the others. An unexpectedly odd size can upset the rhythm of a climb or unnerve the inexperienced.

Electron ladder is best rolled into a small space by twisting the wire between each rung into the form of an X and pulling the rungs tightly over each other. Inspect the ladders frequently, keeping a careful eye out for any fraying of the cable, bending of the rungs, or looseness where they are attached. It is a good practice to clean them thoroughly with a stiff brush after use to remove any superfluous mud, and a preservative like lanolin is good for keeping metal fatigue at bay.

Pitons or metal pegs for hammering into rock crevices are often a club matter. Today there is a wide range of pegs for almost every type of contingency. At all times pegs must be used with respect. Piton hammers also come in various sizes but Cassin makes one specially for cavers with a good long shaft, a square hitting-face and a useful, long chisel-head.

Etriers or short lengths of ladder can occasionally be used underground where a pitch may be too short to waste a standard length or for use in artificial climbing up a difficult pitch in order to fix a standard-length ladder. This is usually only necessary if no maypole or sectional scaling pole is available. A maypole is used for hoisting a ladder into an otherwise inaccessible situation. The clamps must grip the pole tightly and it is often worthwhile using wedges under the clamps to prevent them from slipping. Provision must be made for guying the pole when it is used above twenty-five feet, otherwise it becomes unstable.

Pulleys have various uses as well underground. They can be used on maypoles for hoisting up ladders, belaying from below or assisting in the hoisting of loads. The eyelet type is best because the hook types can slip out.

Another club ' buy ' is an inflatable rubber dinghy. This is useful for crossing underground lakes or making subterranean river journeys. The long narrow ones are best as they are more manoeuvrable and get through tight passages with greater facility. Round ones tend to spin round like crazy coracles. The dinghy should hold two as it is unwise to send one person off ' boating ' alone. For all but the strongest swimmers, life jackets should also be provided when using a dinghy. Make sure that the paddles are fastened to the dinghy itself so that they cannot be swept away. Dinghies also come in handy for ferrying heavy gear and it is a good idea to see that there is a lanyard running all round so that people can hang on to it while working in the water. And do not forget to fasten the pump in with the dinghy equipment as it requires an awful lot of lung power to blow up even the smallest of them. A repair kit is also vital.

Some people make their own dinghies. This can be done with a car or lorry tyre with canvas fastened across the bottom of it and lapped over the sides. But this makes a round ' boat ' which is difficult to steer and there is always the danger of the bottom falling out. In Yugoslavia I have seen a precarious device which looks like a coffin or orange box made out of wood and plastic.

This is not a very safe craft and I cannot recommend it.

Communications underground, apart from the obvious whistle and rope signals, are also a group matter. Field telephones are the simple answer and these can be obtained from any Army surplus store. In spite of their being designed for rugged military field work, they tend to collapse under the wet conditions of caves and every effort should be made to keep them dry. The ideal cave telephone has not yet been designed but experience in the Gouffre Berger has led to some excellent developments in this field pioneered by Dr Harold Lord. He has produced a mine-phone which overcomes many of the disadvantages of the heavier sets. Radio is of little use except perhaps on open shafts, but there is hope in the near future of news of guide-wire radio which will be a boon to cavers generally who use electron ladders and rescue teams in particular.

Clubs should also provide all the ' navvying ' gears required for digging etc. This should include the obvious things like shovels, entrenching tools, crowbars (all sizes), pick axes and the like. Shuttering for making concrete dams, shoring for protection during excavations, trowels, sandbags and other civil-engineering equipment also has its part to play. Pumps to help clear siphons or hold back water while damming should be included in the collection of club tackle. Hose-piping will be required to get the water away. Drag sleds or bogey trucks also come in useful in some of the more determined efforts at digging to clear away debris from the site.

And a final word about explosives. These are in fairly common use underground but this is a subject for experts only. Some is obtained legitimately and some not so legitimately from a friendly quarryman for the price of a pint. I have seen instances where explosives have done more harm than good in a cave—not to mention the possibility of causing serious injury. If blasting has to be done, call in the expert and keep well clear yourself.

4 *Techniques: horizontal*

Quite simply, the art of caving is the art of taking care. It means being able to move safely in an unnatural environment, the realm of unending darkness. And it requires special equipment and skill in using it. The novice caver must learn how to climb on rope and electron ladders, fasten belays, use the appropriate knots and how to crawl, wriggle, climb rocks. He should also be able to swim; some people think swimming is not necessary, but the caver so often comes across water underground that only familiarity with swimming will remove that element of panic which develops with people unused to water in rather more frightening circumstances underground.

The best way to learn all the correct techniques of caving is to join a club or take one of the special caving courses provided at such places as Outdoor Pursuits Centres which have professional as well as teams of voluntary instructors. Most clubs have graduate training courses which educate a budding caver to the right standards of proficiency to the ultimate ability of being able to lead one's own expeditions. Caving is essentially teamwork. It involves working with others and sharing responsibility for each other's welfare and safety. Caving should never be attempted alone or without the proper equipment.

Before going underground it is a good thing to learn something about how caves and potholes are formed. In Britain, most of the known systems are surveyed and described in a number of handy guide books which give the locations of the caves, a brief synopsis of the problems involved and the basic equipment required. As it is not always possible to bring these up-to-date to cope with the speed of new discoveries, it is wise to check through the various club literature if any significant variation has been made in a cave or pothole. The classification system is easy to follow and is arranged in standards of increasing difficulty. As in rock climbing, the grades are something of a compromise and calculated on average ability. But the classifications do give a general guide and form the basis of the assessment of the difficulties to be met.

In the guide books caves are classified by the initials E.C., M.C., D.C., V.D.C., S.C., S.S.C., which stand for Easy, Moderate, Difficult, Very Difficult, Severe and Super Severe. Potholes are similarly classified as E.P., M.P., D.P., V.D.P., S.P., and S.S.P., again ranging from Easy to Super Severe. It is usual for a beginner to be taken through the grades and trained up to tackling the higher standards of cave or pothole. The Super Severes are, naturally, the toughest of the lot and amount to expeditions requiring the utmost endurance and considerable knowledge. Other indications in the guides show whether a cave is of archaeological interest or a show cave, open to the general public and illuminated inside.

It is often best to take a novice into a cave before going down a pothole. A cave involves horizontal movement and it is the best place to get used to the new environment. The easy caves involve nothing more than straightforward walking and keeping one's eyes directed to the ground for obstacles in the way of boulders, pools and holes in the floor. A careful watch must also be kept at head level and above it to avoid unnecessary bumps and bangs and also to miss various pendant formations which may be damaged by a careless blow. It takes time for the human eye to accommodate to darkness and the entry into a cave should be gradual so that the eyes become accustomed to the transition from normal daylight to the more restricted light from one's headlamp. Just inside the zone of darkness, it is a useful tip to close one's eyes for a minute, standing perfectly still.

Once used to the general cave scenery based on a short ramble underground, the other techniques follow naturally. Eventually, the roof lowers and the caver must adopt a bent posture with the head held as horizontally as possible so as to see the roof and the floor at the same time. Be wary of standing on loose boulders, especially on a slope when there are others below. Walking like this can be tiring and some of the strain can be taken by putting the hands on one's knees or thighs to hold some of the weight of the upper body.

Another form of passageway is the 'Monkey Walk' which is not high enough to stand up in and yet not quite low enough to involve crawling. A kind of monkey's shamble taking part of the weight on the hands is best for this situation. With the lowering of the roof, hand and knee crawling is necessary and not for nothing are some passages called 'Knee Wreckers'. Kneeling on sand and soft mud is not too bad but many passages have broken or stony floors and kinds of unpleasant protrusions which can harm that most tender part of the human anatomy, the knee. If it is known that there is to be some lengthy knee work it is as well to wear kneeling pads which protect the knee joint. Some cavers wear them all the time as they give protection against bumps and bangs in other situations as well. Crawling on the knees is tiring work and can be avoided if the crouch position is adopted. But some people because of their size and general physique just have to get down on their knees. For a long section, try and adopt a rhythmic movement moving hand and opposite knee at the same time.

Inevitably, even this method ceases to be of use as rock ceiling and walls get closer together as in a bedding plane or a tube. It is this kind of situation one meets with in caves of the harder classifications. A passage like this is known as a 'squeeze' and that is what it literally is. One has to compress the body in order to get through. The difficulties depend on the size and suppleness of the individual caver. Sometimes it is best to go head first along a tube. In some instances progress is best made lying on one's back digging in with the heels. But it is not always possible to bend the knees to do this. It may be necessary to turn on one's side to cope with the changing shape of the tube. I find it advantageous to extend one arm ahead to reach out and pull on any available projection, using the flat of the other hand to give some pushing traction. One arm ahead also reduces the shoulder width; shoulders and buttocks are the most difficult parts of the anatomy to get through.

At no time try and hurry a passage like this. Threshing and struggling violently is of little use. The best way is to move slowly, inches at a time. Sometimes a kind of ' shimmy ' or serpentine writhing will assist progress. But if you look like getting stuck, don't go any further and retreat while you have chance. For there are some squeezes which some cavers will never be able to manage as big, fat cavers cannot get through where little, thin ones can. It's as simple as that. If the need to force a squeeze is desperate, the last-ditch method is to strip off all one's clothing and proceed by what Norbert Casteret calls ' reptation '—moving almost by skin action. This, of course, is the last resort where the needs of exploration can only be furthered by such a method.

Coping with a tube is one of the most fatiguing aspects of caving and the danger is of getting stuck. This usually happens when a caver panics and struggles violently leading, very often, to his becoming even more firmly fastened. It is also known that under stress, the body gets a certain amount of peripheral expansion, swelling slightly, and this can add to the problem. People also become psychologically stuck. If they have no great experience, they quite fatalistically accept that they are wedged for ever and lose the will to do anything about it. This latter condition can be overcome, generally by talking the person out of it. The thing to do is never to unevaluate the situation because delay can add to the problem. Very often, gentle pulling is sufficient to extricate the wedged caver, particularly if he can be persuaded to become completely relaxed, admittedly a difficult thing to do under the circumstances.

Exhaling, cutting off clothing, pushing and pulling (that is if there is someone on either side of the squeeze) is usually enough

9 Study of a caver, 2,000 feet below ground in the Gouffre Berger

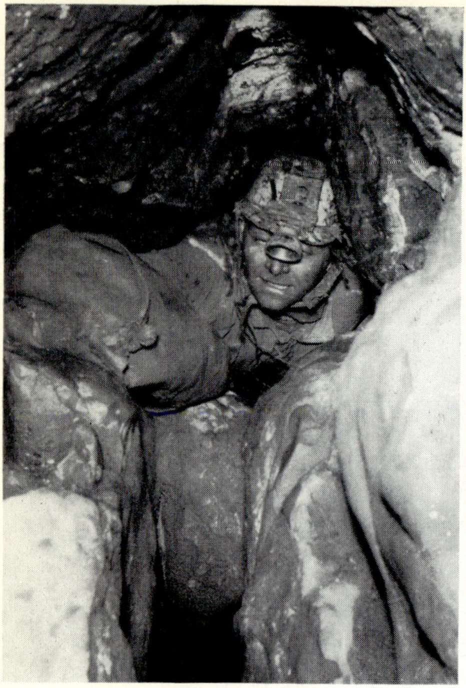

10 *(left)* Typical of the very narrow space in which a skilled caver can manoeuvre. 11 *(right)* Back and knee technique used in climbing a rift or chimney 12 *(below)* The safety line should be kept as clear as possible from the ladder to avoid being trapped underneath it or wound round it

13 Bridging in order to traverse along a meander. Note the plastic water bottle, an important piece of equipment

14 *(above)* another way of traversing using a karabiner clipped to a double rope for extra safety. 15 *(below left)* Caver in action on a wire ladder with a wet cell battery. 16 *(below right)* The right kind of ladder technique, with feet and hands in correct position—but the caver should be wearing a lifeline.

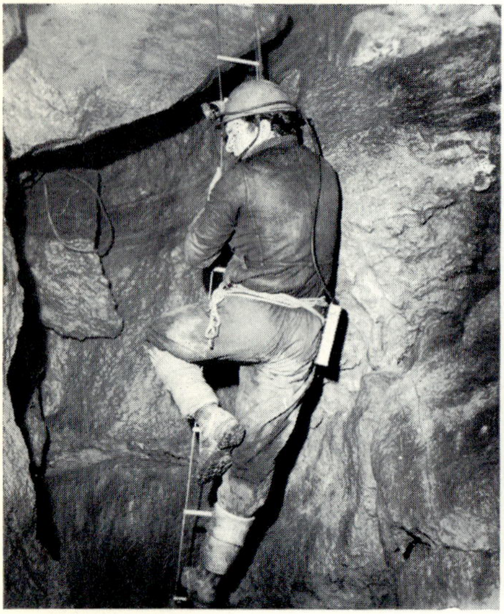

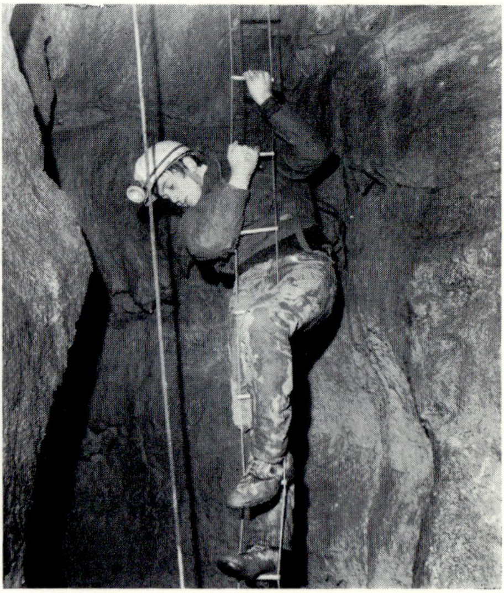

to get out of the predicament. This kind of thing can be avoided if the novice is skilfully trained in getting through tight places until he is physically and mentally prepared to cope with the problem. It sometimes happens, of course, that all members of a party can get through except one. It is better to call off the whole trip rather than leave one person alone on the other side of the squeeze, unless the entrance is visible and there would be no immediate problem of getting out.

In all these lateral movements, personal equipment can be carried slung for part of the time, then either pushed ahead or dragged along on a cord fastened to one's ankle. The pushing method is best because it is not likely to impede progress such as when a bag is being dragged behind catches on something. The position of one's accumulator or pocket battery is also important in tight crawls and this is where the acetylene lamp is an advantage as there are no extraneous wire or accumulators to snarl on rocks.

Another problem to be met with while travelling laterally is that of water. Underground it is met with in the form of waterfalls, lakes, rivers, streams, canals, and sumps. In many instances, water passages can be traversed simply by paddling, wading or swimming. Inflatable exposure suits are a help in some situations and so are rubber dinghies where water depths are known to be too great for walking on the bottom. Whenever water is met with it should be treated with the greatest respect. In all water passages, always keep an eye on the water level and be prepared to retreat at the first signs of any significant rise, unless you are in a cave where there are emergency provisions and facilities for bivouacking or a safe, alternative high-level route. Stream and river passages make exciting caving, but sometimes the roaring of the waters has a numbing effect on the senses. It also makes communication difficult.

Care should be taken when wading along as some cave-stream floors have sizable gulfs full of water and it is easy to disappear down one. Also remember that while it may be easy to walk downstream, the return journey, especially when one is tiring, can sap a lot of energy. If there are traverses out of the water, take them to conserve strength. What to do if cut off by flood is dealt with in Chapter 8.

Cave-diving is also dealt with in a separate chapter (Chapter 7), as it requires specialist equipment and techniques. These are necessary to get beyond a sump where rock and water meet. In some instances the distances under water are quite short and, with training and a certain amount of nerve, they can be dived, using the power of one's own lungs. Some sumps or siphons can be effectively lowered to create an air space by baling them out. It sometimes happens that the water goes back immediately and it may be necessary to do a bit of engineering and build a concrete dam behind which the water which has been baled can be retained. Occasionally it is possible to pump out a siphon and I have done this with rather antiquated hand pumps and with a most sophisticated, electrically powered centrifugal pump. Petrol- or diesel-fuel-driven pumps are not recommended underground because of the danger from exhaust fumes.

Getting through a sump on lung power requires nerve, an ability to swim and a knowledge of the length of the sump involved. At the moment, a sump index is being compiled, which will give the vital statistics of most of the bigger ones which require diving gear. The smaller ones should only be dived if they are known because there is very little margin of safety. It is one thing to dive and swim twenty yards or more under water in a heated public swimming bath which is well lit. It is an entirely different matter to make a much shorter journey in the chill waters of a cave, where the water very often is far from clear, being cloudy with churned-up mud, and where the room to manoeuvre is often restricted. A duck will sometimes only involve taking a deep breath, bending down and bobbing a foot or so under a shelf of rock to an air space on the other side.

For the slightly longer sumps, a rope or guide wire is left *in situ,* and the caver can use it to haul himself through and guide himself in either direction. Sometimes it is a good thing to go through on one's back, pulling on the line, with the head as near to the cave roof as possible, so that the first air space, possibly only just enough for the nose, may be spotted quickly. Take a briefing on the length of the sump beforehand and, if wearing one of the inflatable-type rubber suits, check carefully that all the air is out of it, as this can force the body to the roof of the duck and jam you there. Also check that possible storm-water may have entered or may be entering the cave, thus lengthening the duck and increasing the danger.

A line fastened round the caver in a duck can sometimes be a help, sometimes a hazard. As a last resort, it can be used for hauling back an unconscious or exhausted ducker, but some times if badly controlled it can twine round the caver's legs, restricting movement. A nose clip and goggles sometimes help. A check should also be made that your light will operate underwater. Some accumulator types let water in and performance is affected.

A sump should never be dived alone, i.e. by one member of the party who may have gone on ahead of the others. The most experienced member should go through first and another experienced person should remain behind while the others, less skilled, go through, so that there is help on either side of the sump. At least one member of a party involved in diving a sump of this kind should be capable of giving artificial respiration either by the mouth-to-mouth system or the Silvester method.

All of this may give a highly gloomy view of sump-diving, but it is one of the cave situations in which a crisis can occur and it is best to be completely prepared for it.

5 *Techniques: vertical*

Having dealt with horizontal movements in caves, I come now to vertical movements. Caves lead to potholes and vice versa so that getting up and down safely is as important as travelling horizontally. There are five main methods of getting up and down pothole pitches. The most generally accepted means of making a descent is on a ladder—either the lightweight electron type or the natural fibre and wooden rung type, both of which I have described elsewhere.

The second method, now growing in popularity and widely encouraged in America, is that of making a speedy descent by roping down, called abseiling or rappelling; and returning via the same ropes by prusiking, whereby the caver ascends by using a special sliding knot which can be pushed up the rope, with the feet in stirrups, or by using one of the mechanical contrivances made specially for the purpose.

The third way is to use ordinary rock-climbing methods and the fourth is to use a knotted rope. The fifth and easiest way up and down a shaft involves the use of a winch which may be power-driven or raised by handles or pedals fixed to a frame with a cable drum attached.

The opportunities for winching do not often arise except for descents of big open shafts like Gaping Gill where a fixed winch may be placed for use over a long period. This involves diverting the Fell Beck and it is a major operation to get a gantry in position across the main shaft and involves a considerable amount

There is a 6th, but this tends to be a speedy, one directional, gravity governed movement, where the person who employs the method may well find himself irretrievably dead!

of engineering. A power-driven winch is too heavy for general use and the occasions when one can be used are limited. Very often winches are more trouble than they are worth. Breakdowns in the motor-powered ones can leave one dangling unhappily in space for lengthy periods. Hand- and pedal-operated winches also have their troubles which include cables becoming fouled. For some descents a bosun's chair is attached to the end of a cable on which one sits with some degree of trepidation. On the deeper descents, such as the greater gulfs in France and the Pyrenees, parachute-type harnesses have been devised. These leave little margin for safety should a fault develop as in the case of Marcel Loubens in the Gouffre Pierre St Martin. He fell to his death due to a fault in the harness construction. Since that catastrophe, an unhappy personal experience on the old, rickety winch in Nettle Pot and being stuck for half an hour in the waterfall in the main shaft of Gaping Gill, I have had little faith in winching techniques. But a small winch can be of use for raising and lowering tackle.

Most cavers start their careers by using ladders and I think they will remain in basic use for a long time yet, no matter what other methods are exploited. They provide a simple and generally reliable means of getting up and down and their usage must be mastered at an early stage. The correct rigging of a ladder is the first stage in its use. It must be fixed in such a way that it stays put. There is no room for sloppiness in fixing a ladder as many lives may clearly depend on it. Ladders can be fixed by header ropes to rocks, pitons, boulders (if they are large and heavy enough), short lengths of crowbar driven into the rock for the purpose, Rawlbolts (expansion bolts), beams or (as I have used successfully) a short length of railway line wedged firmly over the top of a shaft. The leader must assess which is the safest location for the fixing of the ladder. Some caves have permanently fixed ladder belays but these should be carefully examined before use to see if they have become worn, loose or inadequate in any way. Even if they are only slightly suspect, the caver would be well advised to find a safer alternative.

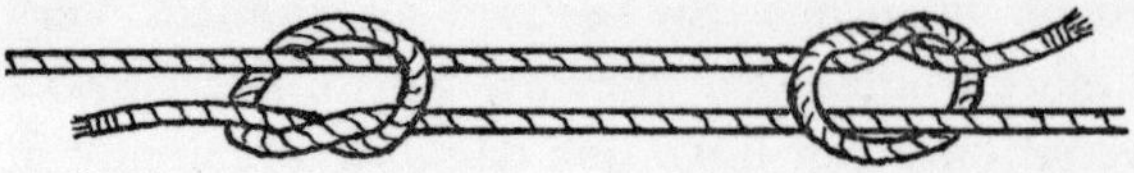

5

6

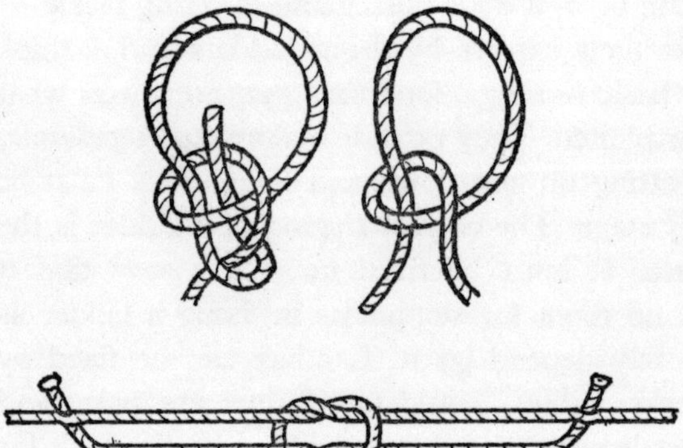

7

8

5 Fisherman's knot for joining two ropes.
6 Prusik knot.
7 Waistband tie.
8 Tarbuck knot.

The leader must also assess, if he does not know already, how much ladder is needed for a particular pitch, and decide on how many lengths must be clipped or fastened together. Care must be taken while this laddering operation is going on as it often takes place at the edge of a gulf and a careless move or an accidental trip over the ladder being prepared for use can lead to the speediest descent possible at the usual rate of acceleration of a falling body of 32.2 feet per second per second. It is safer, in fact, to tie on to somewhere safe while setting the ladder in position as it involves a fair amount of leaning over and peering into the void and jerking movements to settle it in position. The great thing, when dropping it over the edge, is to make sure that it is firmly fixed on the belay point. I know of one club which had to go and borrow another club's ladders to retrieve their own, which had been carefully lobbed over the edge when the man responsible for fixing it, had failed to do so.

Rope ladders are best secured to the belaying header rope by the normal method of joining two ropes together, i.e. the fisherman's knot or two thumb knots, or by a reef knot with two half-hitches. If the ladder is likely to be there for any length of time, the ends can be stopped to the header ropes with a twine whipping to make them more secure. The thumb knots must also fit snugly into each other.

Electron ladders are easier to fix as they can be fastened by the C rings to specially made wire tethers. So as not to cause damage if a stalagmite boss is used for belaying, a leather sleeve or piece of sacking should be used to stop the wire cutting into the formation. These lightweight ladders can also be fixed direct from a piton or an expansion bolt and must only be used when absolutely sure that it has a safe anchorage and will not slip out, no matter what direction the pull comes from. Joining extra lengths of electron ladder is also easily effected by using the C rings, but rope ladders must have the 'tails' or trailing ends securely tied to the headers of the next length.

It takes experience to get the 'feel' of a correctly rigged

ladder. Every effort should be made to get the twists out of a ladder as it is suspended down a shaft because a twist, if it is tight against the rock wall, can be difficult to negotiate. It does not matter quite as much if the ladder hangs freely away from a shaft wall. Holding the arms like a derrick and shaking the ladder often help to settle it into the correct vertical position without twists. An experienced caver should also be the first down the ladder, securely protected by a lifeline, to sort out any twists or free the ladder from any other snags it may have hit while being unrolled and lowered down a pitch. All too often, lifelining is taken too casually and far too many risks are taken in many instances. The lifeline may be one of the several ropes I mention elsewhere. Whichever is used, it must be securely fixed so that the person controlling the descent is firmly anchored so that a fall will not pull him off his stance. A fellow-member of my own climbing club, Ken Tarbuck, has evolved the knot which bears his name which can be used for tying the lifeline to a waist-band karabiner (it must be screwgated for extra safety) for a caver on the ladder or the lifeliner's anchoring rope.

The Tarbuck knot slides and grips on the main ropes under the combined effects of velocity and high tension and acts as a shock absorber eliminating the sudden jerk which can cause ropes to break. But it is a knot which requires the greatest care in tying; in particular the figure-eight finishing lock must be correct (*see* fig 8) or the rope can run and untie itself. Using the knot for anchoring the lifeliner, the active rope must be taken from the waist over the rock bollard or belay point and back to the waist. Pass a bight of rope up through the waist-band from below, and tie with an anchor knot, making certain that the active rope, which will be hanging down, is brought up parallel with the anchor rope and embraced in the figure-eight knot. Length adjustment can be made by sliding the Tarbuck knot, and it should be just taut when the stance is taken up by lifelining. Sometimes, the only available anchor point may be too small or restricted to allow a safe lodgement of the climbing rope, so an independent sling of thinner nylon rope can be used and clipped into the waist karabiner. If it is too short to allow this, the active rope must be taken through the karabiner in the sling and brought back and tied to the waist in the normal manner.

Quite clearly, the lifeliner must not be pulled from his position and the stance taken must not be casual. Sometimes, on a short pitch, it may be adequate for the lifeliner to jam into position or brace his feet against an opposing wall or boulder, but I still feel it is safest to be tied, so that a sudden fall cannot bring disaster. The stance should be so that the rope passes round the body with one hand controlling the uncoiling rope and the other directing it down the shaft. Preferably one hip should be braced forward and the lifeline passed above the anchor ropes with one complete turn round the controlling forearm (from inside to outside—never under any circumstances on the directing arm). The controlling forearm is braced against the thigh and the braking is done with that hand and arm. Rotating the wrist on the controlling forearm and the pressure of the hand are part of the braking system and further energy and shock is absorbed in the body, again reducing the sudden jerk. This rope work with the appropriate knots should be taught and practised to perfection before a novice goes underground.

Also before a novice makes a descent, he should be taught the correct whistle and rope signals which act as a guide to the lifeliner when he may not be able to hear shouted instructions. One blast or one pull on the rope means—Stop paying out the rope. Two blasts or two tugs on the rope means—Take in the rope. Three blasts or three pulls means—Continue to pay out the rope. One long blast or a long, continuous pull means—An emergency, hold tight. A sensitive lifeliner can often feel whether the person on the ladder is in trouble. Care should be taken not to accumulate slack so that if the person on the ladder falls, it comes with a jerk on the rope and then on the lifeliner. A lifeliner can often gauge the rhythm of a descent or a climb and pay out rope or take it in accordingly.

9

10

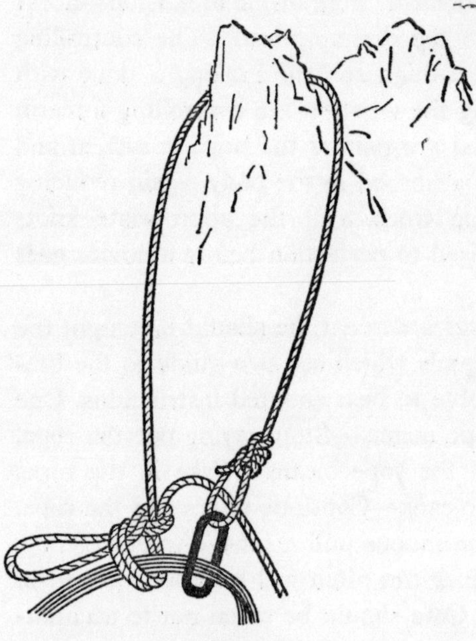

11

9 Braced position of the body friction brake to arrest a falling body.
10 The Tarbuck method of using the body friction brake.
11 The Tarbuck method of anchoring on to a belay with waist sling and the active rope.

For the actual descent on the ladder, two methods are used. On rope ladders with wooden staves, the simple way is to climb on it with both feet from the front, pushing them on to the rungs from the front and gripping the rope sides of the ladder from behind with the hands crossed over each other. This helps to keep the body upright. The main thing is to keep as upright as possible, otherwise the tendency is for the feet to push the ladder away, the strain (which is considerable) comes on one's hands and forearms quickly causing tiredness, and a large area of the body leans backwards over the gulf on which gravity can take its toll.

On the electron ladder the best technique is to make love to it. That is, to cross the hands behind and grip the rungs, one hand over the other, and to have one foot on the front of the ladder seeking purchase from the front and the other round the back of the ladder with the heel slotting on to the rung from behind. A doctor-caver friend of mine says this is an uncomfortable way of making love, but this is how he describes the correct method of using an electron ladder.

This applies of course where the ladder is hanging freely in space down a shaft. When it is close against a rock wall, both toes must enter the rung from the front and a kicking movement of the foot backs the ladder away from the wall. If the foot is pushed right up to the instep, this holds the next rung away from the rock so that the other foot can be inserted in it. Try always to get the foot in as far as the instep in such a position that it reduces the risk of cramp in the calf and slipping if only nebulous toe-holds are used.

Going over the edge where the ladder is tight against the rock is another problem, and it is as well to have some of the strain taken on a tight rope from the lifeliner while you pull the ladder away in order to get a purchase. At this stage, the rungs can be held from the front until there is sufficient free space to adopt the behind-hand-and-foot technique.

Frequent practice either from a beam in a gymnasium or the branch of a tree or against the side wall of some building or outside rock face helps to give the confidence so necessary for ladder work. Remember always to try and keep the body as vertical as possible. Many novice cavers, like novice climbers, try to climb entirely with their hands and arms, which is exceedingly tiring and generally useless. Keeping the body upright spreads the load through the great lifting muscles of the thighs and small of the back. By distributing the weight and lifting or descending load in this fashion, more muscle groups are brought into play and fatigue is greatly diminished.

Seek an even rhythm of movement and try not to move in a series of tense jerks. The perfect caver ' flows ' up and down a ladder but it takes practice to pass from the first gentle struggle on a twenty-feet pitch to the 300-feet main shaft of Gaping Gill.

Do not hesitate to rest if you tire. Sometimes it is possible to step off the ladder on to a convenient ledge where you can get a second wind and build up steam for another section of the climb. If there is no ledge, clip on to the ladder with the karabiner and this is where the chest harness or rope baudrier is a boon as this keeps the centre of gravity higher and causes less strain if the caver has to let go and just dangle. The waist-loop tends to crush up the bottom of the rib cage in such circumstances and also increases the possibility of toppling upside down.

The greatest strain naturally comes when ascending the ladder, and the lifeliner must sense when a good strong pull is necessary. Care must be taken not to tug too hard as this can upset the rhythm of a climber or pull him off altogether. If a man is too exhausted to climb, a separate rope may be used—if he is not too far gone to assist his upward journey with, if possible, a hauling party separate from the lifeliner who must continue his function.

Incidentals to ladder work include the careful working-out of a programme. Sometimes, a man may be left in a ' sentry-box ' or some other convenient belay point to await the return of the others in the party and assist and protect their ascent. This is a long, boring job for somebody and it is often preferable for all the party to go down. To do this safely, the lifeline is threaded through a pulley preferably from the ladder anchor point. The lifeline passes through the pulley block and the lifeliner can tie on and be lifelined from the bottom of the pitch or pothole by a second man who must also be secured by the basic anchor technique. Every care must be taken that when the last man down reaches bottom, the two ends of the lifeline be joined together so as to form a continuous loop. Losing the end which may whip up through the pulley could mean that someone would have to climb the ladder unprotected to re-establish the belay. Some have done this but it is bad practice. The rope through the pulley block must be fixed at the bottom of the pitch so that it may be used for lifelining the first man back up. If the expedition is big enough it is best to let two or three of the more experienced go up first with an experienced person left to the last. In this way, the less expert going up in between can be given every assistance as a man below the ladder can see many of the snags which a man on the ladder may be running into.

Apart from getting cavers up and down shafts, ladders may be used in various ways, such as making an aerial tightrope above an awkward waterfall. This can be done by suspending a section of ladder between one fixed point and another horizontally and balancing along it using another rope as a hand rail and a lifeline for any emergency. But inevitably, the laddering of a waterfall cannot be avoided. This requires no difference in technique, only greater care, because the water pressure can sweep a climber off a ladder and the unending drumming sound can have a bemusing effect.

Concluding the subject of ladders, the final problem of all is getting them back out of a shaft. This can be done in a number of ways. The last man up the shaft can tie a spare line to the end of the ladder and it can be brought up while he is climbing on a higher section of it so that he is on hand to guide it round any snags which may occur. Or he can fasten it to his own waist-band or a short length of line, trailing it behind him as he ascends.

In the caving world at the moment, there is some controversy about what are called 'vertical techniques', through which quick descents are made by abseiling and ascents by prusiking. Abseiling is a method of roping down; this has been acceptable practice for a long time with the ascent being made by ladder. But the prusik method is coming increasingly into its own. Prusiking was first developed on the Continent mainly as a means of getting a climber out of a crevasse. It involves the use of sling stirrups into which the feet are placed to keep the body upright and reduce tension from a waist-band while a prusik knot or friction hitch is looped over the lifeline. The method of ascent is to slide the knots alternately up the rope, transferring the weight from one stirrup to the other. A third sling round the chest was also brought into use which further kept the body upright.

As far as caving is concerned, this abseil-prusik method is winning a growing band of adherents. Some people may think it only satisfactory for short distances where a quick in-and-out is necessary, but an American team has used this two-way method in the Sotano de las Golondrinas in Mexico, a free descent of 1,094 feet, and the even deeper Sotano de Tlamaya (1,400 feet) has been regularly bottomed and climbed in this fashion. Supporters of the method claim that it is safer, because increased speed reduces fatigue, and that it also enables more work to be done underground. Its critics point to the occasional accident (few compared with other forms of accident) and stress that prusiking provides fewer safeguards, in so far as there is less opportunity for belaying and a single rope has been known to break. As I have only used this method in an emergency while climbing, I feel unqualified to speak but must mention it as one of the newer techniques which may eventually revolutionise caving methods. Gadgets for ascending which lock on to the climbing rope have been evolved for rock-climbing and are now also in use underground. (See Chapter 2.)

With abseiling I am on safer ground and, for the experienced, it is a quick method of descent obviating the slower and more tiring ladder work. There is less security, however, as lifelines are not normally used in conjunction with abseiling and considerable training is necessary before entrusting oneself to this method of descent.

There are various ways of abseiling. The oldest method is to use a double rope from a belay and drop it over the edge of a pitch. Every effort must be made to ensure that it reaches the end of the pitch, otherwise you are in trouble if you reach the end of the rope before the end of the pitch. Roping down this way, the caver faces inwards and stands astride the double rope. The rope should pass inside the right thigh, up the outside of the thigh and across the chest, over the left shoulder and back to the right thigh. The rope is gripped in the right hand about waist level and the left hand holds it about level with the face.

This method is useful for quick, short descents. Longer descents may be made with a variation of this classic method by passing the rope round the left thigh, up in front over the left shoulder, round the back of the neck (making sure it is protected by a turned-up collar, but never a scarf as this can be tightened round the throat if caught up in the rope) and again to the front over the right shoulder. The right hand in each case is the brake.

Much more secure is the snap-ring method using a karabiner and sling. The sling is formed in a figure eight and the legs thrust through the loops. The centre of the 'eight' is pulled up into the crotch and a karabiner clipped over it. The double rope is then threaded into the karabiner which should be firmly locked so that it will not spring open. Leaning out backwards, the caver can then either walk down the pitch or, if it is a 'free' pitch, slide down the rope under the control of the hands, keeping the knees together and breaking when necessary.

As gadgetry is coming more and more into caving, some use the patent descendeurs. The Fisher descendeur is really two steel loops: the smaller one which is only an 'eye' is clipped to crotch-sling and waist-loop, again locking the gate; the double

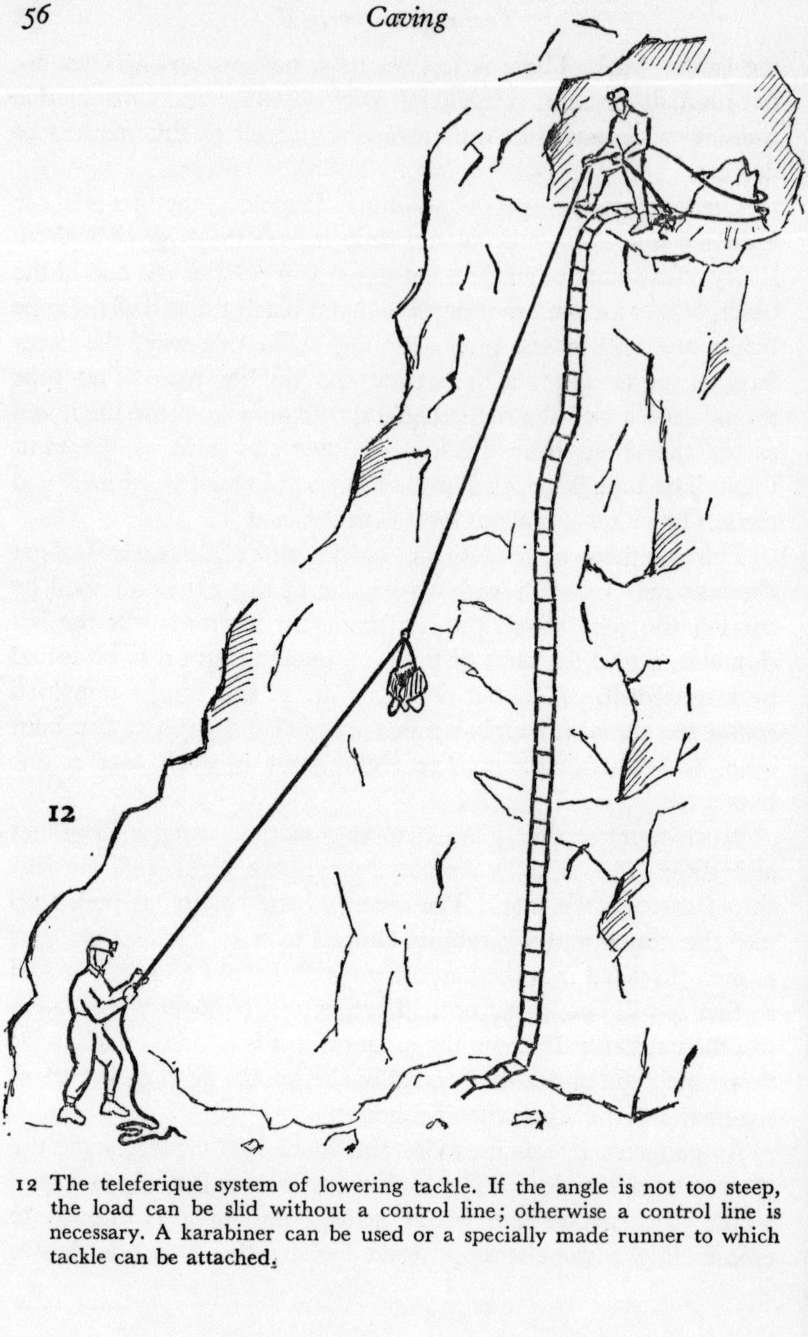

12

12 The teleferique system of lowering tackle. If the angle is not too steep, the load can be slid without a control line; otherwise a control line is necessary. A karabiner can be used or a specially made runner to which tackle can be attached.

rope is looped through the bigger 'eye' and round the back of the smaller one. One hand is held above the descendeur and the other below it. By passing the free part of the abseil rope between the top 'eye' and the rope from the top, this will bind on the rope and lock it if it is necessary to make a prolonged stop. It needs a tug to free it to continue the descent.

The trident descendeur is an alloy fork, light in weight, which also has a small 'eye' for clipping to crotch and waist band. The abseil rope is curled twice round the descendeur prongs and brought up through the curved-over prong. During the descent, the left hand holds the descendeur itself and the right pays out and controls the rope.

In some places, knotted rope is useful as a hand rail and for getting up awkward, slippery slopes and the shorter pitches where there may be adequate footholds which have become too slippery for the hands. The knots are best placed about two feet apart and are necessary to prevent the hands, often wet and coated with mud, from slipping.

The other basic method of getting up and down is that of straightforward rock-climbing but invariably, due to the limited vision one usually has underground, this cannot be done to any great extent. Some dry pitches of steep wall can be climbed direct if the way is known but dangers arise from boots becoming coated with greasy mud and the possibility of a limestone block coming out. The more general climbing techniques used in caving are those of bridging, chimneying and hand-and-foot traversing. Chimneys are met with frequently underground. Narrow ones can be climbed by jamming with knees, feet and hands. Much depends on the width of the fissure. Some can be climbed by straddling with the feet on opposite walls with the hands seeking holds on each side. If narrower, the chimney can be climbed by wedging one's backside against the wall and putting one's feet on the opposing wall, climbing by means of sliding the buttocks upwards and jamming across with the hands while the feet are brought up to a new position. Another way is to press the palms on the opposing walls with one foot thrust forwards and the other bent backwards. By pressure with the palms and a thrusting movement from the bent leg, using the other to seek out holds and jam, upward progress can be made. Straddling technique is used in high meanders and it is best to pick your own line of traverse most comfortable to your leg width. In these movements it is not always necessary to have a lifeline except in the cases of the most inexperienced because, in a chimney, the dangers of falling are not too great unless it dramatically widens beyond the span of hands and legs.

In any event, all these manoeuvres should be taught and practised above ground, preferably on limestone, so that their use underground, often in dark and wet conditions, becomes second nature.

6 *Other aspects of caving*

I have covered some of the aspects of getting up and down and across. Other means are also used which are incidental to the others, designed to cope with the odd problems which occur underground. One of them is how to get up a sheer wall. It often happens that further exploration can be made along passages high above the one you may be in but there may be no rock-climbing or other feasible route to reach it. Some cavers are also brilliant rock-climbers and it often happens that there is a 'tiger' in the team who can at least reach the higher point by conventional methods or artificial methods with pegs, etriers, etc. Where this cannot be done, the scaling pole or maypole is brought into service.

The reason for its name is obvious when it is festooned with ladder and ropes. A maypole is made up in tubular sections which are clamped together until they are long enough to reach the required height. At the top end, a special clamp is fixed or it may have the top of a T piece welded on, to which can be fixed the ladder, a pulley with the lifeline threaded through and two guy lines for bracing. The pole is set in position at any angle and the top end must extend three or four feet above the point to be reached to facilitate getting off the erected ladder. The pole should have its base firmly held by boulders all round to prevent movement, and further security can be given to prevent the pole being pushed away by a line connecting it to a bollard or piton on the wall to be climbed.

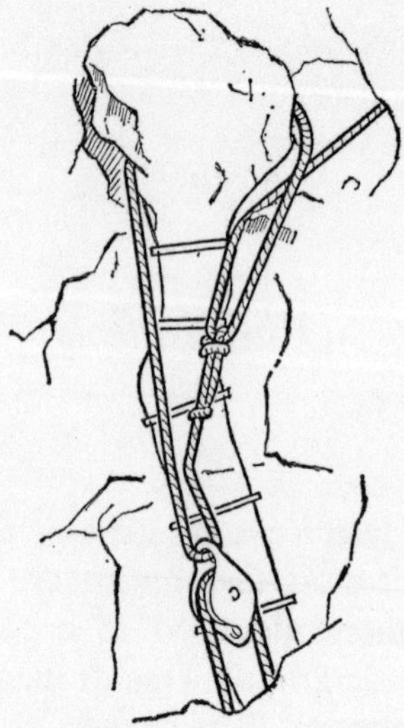

13

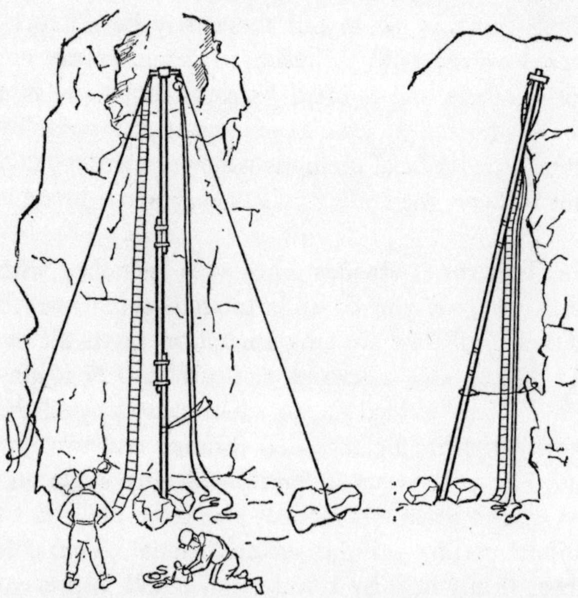

14

13 Pulley block technique for lifelining when last man down or up must be belayed or protected from below.

14 Scaling pole in position for hoisting ladder up unclimbable pitch. Note clamps, angle of lean and pulley block.

Once the first ascent has been made the ladder is fixed in the usual way and the pole can be brought up and de-sectionalised if further exploration beyond is contemplated. The belaying, of course, is done from below via the rope threaded through the pulley mentioned and the lifeliner must be secured. In addition to the bracing of the guy lines, the pole may also be held in position by members of the party to prevent lateral slide.

It is often necessary underground to make extensive traverses. The more usual ones are made by walking along narrow ledges, sometimes over a pitch or along one wall of a passage which may be too wide to bridge or straddle. A traverse can also be made to avoid a stream passage or a lake. Ledges are found which are satisfyingly large and it is no more difficult than walking along a pavement. But it may narrow or the traverse may become a matter of stretching in balance from footholds. Low down, there is little to fear from a slip unless the water below is excessively deep. But high-level traverses often going into the roof of a cave must be protected. This can be done with pitons driven into the rock, with slings and karabiners or, if the traverse is often used, with a fixed rope or wire which can be be used for the hands or for clipping on to.

The insertion of pegs should never be left to the novices in the party. Picking the right peg for the right type of job is important. The pitons should be carefully spaced out along the traverse and put in horizontally or, preferably, at a slight angle downwards and never pointing upwards. A piton should be driven in with the eye hole or ring as close to the rock face as possible to give maximum purchase. If there is too much extrusion of the peg from the rock, a dangerous amount of leverage can be exerted in the event of a slip if the eyelet at the end of the peg is used. In these conditions it is far safer to use some of the new climbing webbing in the form of a tie-off or 'hero' loop right up against the rock face, with a karabiner through the loop.

Webbing made from nylon, first used in America, is also coming into use underground for making into etriers as well as providing protection. In many circumstances, it is better than rope, e.g. when tied over a peg, as its breadth gives it a better grip. It is also useful for making gear-carrying slings and also for belays.

Limestone is an unstable rock and must be treated gingerly because of its nature; earth movements, erosion, distant explosions from quarrying, even shouting and whistling can bring down a fall of rock. You may dislodge rocks yourself and if you are above someone, always use the warning cry ' Below '.

Rock falls often choke up cave passages with boulders which may lie poised over other boulders and drops. Only one person at a time should move on a boulder slope, unless individual parallel lines can be taken. Too many people on a pile of boulders can cause a rock avalanche which may block other passages. Balance must be well maintained and the feet placed firmly down to avoid trapping or falling and breaking an ankle. If it is necessary to crawl through a boulder slope, struggle as little as possible and keep away from the sides to avoid further rock movement. Routes across some of the boulder chokes and slopes are often clearly defined with use. A splash of white or luminous paint can be a help in some circumstances or a long-burning candle if the return by the same route is to be made within a reasonable time.

Loose rocks are often met with in mines. Although man-made, mines can be as complex as any caves. The great thing is not to use anything left behind in a mine on trust. Rock walls built up by miners may be safe, they may not be. Treat them as if they are not safe. Nor should you use the wooden stemples, iron spikes or winches left behind by the miners of old as it is impossible to tell how safe they are. Some shafts are lined with stones, magnificent examples of masonry, but they are often unreliable for hand- and foot-holds and should be treated accordingly. The roofs of artificially made passages are also suspect and so are the props and baulks of timber left to shore them up. Water and general decay wreak great havoc on the equipment of long ago as it has often been neglected for centuries. Do not be tempted to ride in old trucks or bogeys left underground as you may not be able to stop them and you may not always know where they are going. Often they run on to some drop where their original contents were tipped out as mine debris—and you may follow them down.

Unlike caves, mines sometimes become contaminated with sewage and there have been recorded instances of illnesses following contact with it in Derbyshire mines. Some mines are also inhabited by rats and while I know of no one being attacked by them, I do know that a caver has been infected with Weil's disease, an infectious jaundice transmitted by rats and which occasionally infects sewer workers. Consequently, after the descent of a mine where there is the least suspicion of sewage or rat contamination, be extremely scrupulous in cleaning up afterwards, treating any cuts and abrasions with strong antiseptics.

Another problem likely to be met in mines is the occasional pocket of carbon dioxide, which has no smell. It can be fatal, but the only time I came across it was in Derbyshire. It made me ill but its presence was quickly suspected and the party I was with was able to get away quickly into a fresh air stream which brought about speedy recovery.

It is in mines that acetylene lamps are most useful, for a naked flame is quickly affected by the presence of the gas. Its physical manifestations include a violent headache, giddiness and weakness with feelings of nausea. The weight of the gas, being heavier than air, tends to make it gather at the foot of a shaft or low down in an airless part of a passage. Before venturing into a shaft where the gas may possibly have accumulated, a candle or carbide lamp should be lowered to check for the presence of gas. It may occasionally be met while excavating long-sealed-off passages but I have no record of anyone suffering lasting effects from a brief contact with the gas underground.

It will be a long time before a beginner is allowed to take part in route-finding. On the first descents into caves and potholes, all attention is focused on techniques and getting used to the unfamiliar underground scene. It takes time and experience for a caver to learn his way around underground, to recognise the signs, landmarks and all other indications of the way his route should go. On many caving trips, of course, it is possible to get a good theoretical view of a cave from one of the many maps and surveys that are made. These, in conjunction with the guide books to the various areas, give a preliminary appreciation of what is to come. The known and explored caves have well-defined routes and they are as clear as routes on an exterior rock face. On each trip study the landmarks, which eventually become as familiar as the tors on a moor. At first, in the general darkness, all caves and rocks look alike but all have their individual characteristics. But some people get to know only one or two systems which they know like the backs of their hands—and get lost when they move into somewhere new. Get into as many different kinds of caves and potholes as you can and study each carefully. The lessons learned from this can be of inestimable help when moving into the realm of exploration of new ways, new caves and potholes which are among the great joys of caving.

In time you will recognise certain things—where to look for a high passage which may traverse above a choke or a squeeze, where a by-pass route may be found in times of flood. Stalactites, stalagmites and other formations which may all look alike at first assume individuality and they become the signposts and marker buoys of your journeys underground. And you will appreciate how some of them get the names they have.

It is only when you have this confidence in yourself, your equipment and your friends that the hazards of caving on which which I have dwelt, perhaps rather heavily, get into perspective, and a cave will then be no more dangerous than crossing a busy street.

The needs of route-finding lead to the creation of maps of the underworld and these are important both for your own benefit and for passing on information to others. Some cavers are jealous of their finds and guard them closely but I believe in the free dissemination of information among the organised clubs. There are fools who go underground, who commit damage, spoil the beauty of caves by taking out souvenirs or leaving rubbish behind. But by and large the serious-minded cavers are in the majority and they have a true respect of the underworld scene and of the efforts of others.

Cave-surveying is, in fact, one of the most difficult aspects of speleology. There are only a few competent cave-surveyors in the country and some of these have come into caving via their own mining experiences. As caves and potholes are classified according to their severity, so are cave surveys according to their degree of detail and the equipment required for carrying them out. The representations can show a cave in section, profile and by exploded diagrams relating the system to the surrounding countryside and geology. The Cave Research Group has seven grades of survey which range from the extremely simple based on rough sketches either from memory or while on the site itself to the thoroughly professional jobs which require expensive surveying equipment. Smaller and less ambitious drawings can be made with simpler equipment such as a prismatic compass and a long tape; but the higher standard of survey requires a surveyor's theodolite for levels, steel tapes, chains, tachymeters and a miner's dial which is useful where a theodolite is impracticable.

Surveying is fascinating, for apart from the time spent on measurements of angles and distances, it gives a full opportunity to study the shape and configuration of a cave and every minute detail of its make-up. Even though you may not have the skills to make a competent surveyor, there is a lot to be said for helping. At any rate, you should keep a notebook of your own in which to make your own rough sketches, measurements and other details which will be of value in your own and other people's explorations. These will be of more value if you learn and understand the various symbols of cave-surveying both to interpret the surveys of others and to delineate your own. I myself have never gone beyond the sketch-book stage, but I have a number of interesting logs of my own compiling with informative, if not accurate, sketches and information about known caves and digs in search of new ones.

Mention of 'digs' has nothing to do with boarding-house accommodation—it is a short form of referring to digging for new caves. It is in this direction that true exploration lies, either inside a known cave or outside. A small depression or hollow in the surface in an area of limestone may lead to immense possibilities underground. A slight draught at the end of an apparently sealed-off cave passage may, with digging and probing, reveal a way into a new part of a system and magnificent prospects of previously unseen formations and other delights of the true explorer.

On my heart, when I die, they will find the phrase 'Snelslow must go'. It is the scene of a dig in Derbyshire, near Giants Hole, where I have gone year after year, hoping that it would break through possibly into Giants and the other systems below Rushup Edge. As I write I have a water-stained notebook compiled by a friend and myself which records the hours and hours of digging, damming (and a little blasting which got us into trouble) and other backbreaking work to force this passage. Yet I have never begrudged any of the arduous hours of digging out layer after layer of mud and cave debris which had to be dragged to the entrance on a simple sled we designed for the purpose. We built a small concrete dam to keep down the level of the water going into the cave—but this was reduced to a farce one day when a cloudburst monsooned on Derbyshire and the little valley was flooded to a depth of thirty feet, completely covering the cave dig entrance. Sometimes we won the battle of the silt getting back in and a bit more of the tight downward path would be revealed. A few weeks later it was back to Square One as the remorseless tide of Nature took back what we had dug out.

Enthusiasm wanes when one can only make spasmodic visits but patience begets its own rewards and caving history is full of stories of those who have gone again and again to the same site and by luck, skill—or even a minor miracle—have found the way.

If the possibilities of breaking through are good, care must be taken immediately. The clay at your feet may be just a blockage in a high passage and you could drop through, so it is safer to tie on in such positions. It is also advisable not to rush on immediately a breakthrough is achieved but to take a slow, deliberate appraisal of the way ahead. Care must be taken not to damage formations or disturb the floor too much as there may be many small items for the scientific caver to find.

If it is an external dig and reveals a vertical shaft or even a horizontal passage into which domestic animals from farms may stray it is important to notify the farmer or landowner and take precautions to prevent this from happening. Digs must also be undertaken with the permission of the landowner and agreement obtained if blasting or other civil-engineering-type work is required. It is also good caving manners to respect the digs of others. There is a gentleman's agreement which is unfortunately breached more often than it is observed, that cavers respect the digs of others and do not poach. Personally, I do not mind who carries on any dig I have been involved in. There may be a certain amount of glory in it but cavers should rise above petty possessiveness. To share the joys of discovery, the new revelations of the wonders of Britain's underground is, I believe, one of the purposes of organised speleology. This ' It's mine, all mine ' attitude is all wrong, I feel—although I agree that some protection is also necessary from Castleton Cowboy Cavers who infest the caving centres of Britain and do much to damage the image of the sport in rural communities and bring speleology into disrepute.

Cave-diving

The advice of Mr Punch for those about to marry—'Don't'—is equally applicable to that esoteric aspect of speleology, cave-diving. Unless one is prepared to devote oneself to the most severe course of training it is possible to devise, it is no use even thinking of taking up cave-diving. It demands the highest mental and physical disciplines. For without doubt, cave-diving is by far the most difficult and hazardous aspect of cave science. It demands a peak of physical fitness and an ability to keep one's self-control under all kinds of circumstances. It is just not enough to be a competent underwater swimmer for, even with the most modern equipment, emergencies arise in cave waters which are special to their *locale*.

Today's modern equipment, of course, bears little resemblance to the primitive tackle first used so long ago when air had to be pumped manually to a diver. Little was known then of the effects of water pressure, the complications arising from an over-rapid ascent from considerable depths and other physiological problems. Yet it was with primitive, manually supplied air that the first recorded cave dive was made in France in 1878—without much success.

In caving, it has always been known that diving in one form or another would be the only way to solve some of the ultimate cave problems of the twentieth century. In many cases, exploration had to cease where rock and water met at a sump which could not be swum through because of the limitations of human lung power. Early explorers were forced to stop at such obstacles but there were some, the foolhardy and the brave, who forced themselves to the limits of human endurance to get through simply by holding their breath. Foolish was the man mentioned in William Bray's historical tour of Derbyshire in 1777; possibly inspired by stronger waters than those from the local springs, he tried to swim through the 'River Styx' at the Buxton Water Rising in Peak Cavern. It reaches a total siphon at this point and this 'hero' tried to dive it into the cavern beyond. In the process he knocked himself out and had to be dragged out more dead than alive. This siphon did not give up its secrets until August 1947, when the Derbyshire section of the Cave Diving Group showed a totally submerged passage upstream to a Torricellian Chamber. Further explorations by divers got them through Buxton Water and Speedwell Water and to a whole wealth of new finds.

The first great breakthrough in cave-diving which showed what could be done on lung power alone, coupled with courage and confidence, was Norbert Casteret's penetration of the siphon in the Montespan Grotto in August 1922. When I met him in Blackpool a few years ago, he told me it was one of the hardest things he had ever undertaken but he was convinced that he could get through. In his own words, he decided to 'dare the underground river and push into the bowels of the mountain'. He made the dive stark naked, carrying his matches and candles in his rubber bathing helmet, to keep them dry. And so he dived the siphon which eventually led to the discovery, with Henri Godin, of that huge treasure house of rock engravings, clay statues and other prehistorical remains. This dive encouraged others to rethink on the psychological barriers to their own explorations and more determined efforts were made to get through hitherto hopeless sumps. The alternative method of getting through was to pump or bale these siphons dry or, where possible, build dams and divert the waters elsewhere to clear an air space.

It often happens that war is the spur to development and invention, and this was no less true with regard to cave-diving. The needs of submarine warfare helped the evolution of frogmen's equipment and experience gained during the war led to great strides in underwater cave-exploring. Before this came about, cave-diving in Britain can be said to have its origins in the work of two Northern cavers, Graham Balcombe and Jack Sheppard, in Swildon's Hole in the Mendips. They first attempted to get through Sump No. 1 using Heath Robinson-style equipment consisting of, among other things, cycle-frame tubing and a length of garden hosepipe. This was in February 1934 and the attempt was not wholly successful. But sufficient was done to convince Balcombe that there was a way through. In 1936, using 'improved' but also home-made equipment, Sheppard dived through the sump and saw enough to convince him that at that point, anyway, it could be dived on lung power alone.

After the war, cave-diving became more organised and received further impetus from Dr Bob Leakey's efforts in the canal in Disappointment Pot in 1944 which led to its eventual link-up with Hensler's Passage in Gaping Gill. The late Donald Coase, another cave-diving pioneer, made further penetrations of Stoke Lane Slocker possible and great work was done by the diving group from Sheffield University as well as by a few dedicated, but independent teams. Unfortunately, cave-diving sustained a great setback in 1949 when James Gordon Marriott, a former Manchester man and a Royal Marine Commando who became a frogman, died in Wookey Hole. He was a guest diver with the Cave Diving Group and his death was not in any way due to any deficiencies on their part or of himself. In many fields of activity it often happens that a piece of equipment fails or develops a fault: that is what happened in Marriott's case. There was a fault, which he could not possibly have foreseen, on the gauge of his oxygen cylinder, and it caused his death from drowning in spite of heroic attempts to find him and resuscitate him.

As Dr Ken Pearce, one of the leading modern exponents of cave-diving, has pointed out, a fatality like this can cause a serious setback to this branch of cave science. As he has pointed out: 'It is a matter of historical fact that after each of the cave-diving deaths in the British Isles, active cave-diving has virtually ceased for a number of years and progress has been severely retarded.'

This is unfortunately true. Apart from the Marriott disaster there have been three others. Jack Waddon was drowned in the Minneries Pool at Priddy in 1962 during diving practice. Garry Gilliatt, a twenty-four-year-old Rhodesian living and working in Britain, died during a dive in Ecton Mine on the fringe of the Peak District in March 1963. He was a member of a sub-aqua group and was not a specialist in cave-diving and is thought to have been overcome by an attack of the 'bends', which attacks divers coming up from the depths too quickly. The most tragic loss of all was one of the country's most experienced cave-divers, capable in all branches of the sport, who was drowned in Lancaster Pot in March 1964. For some reason his mouthpiece came adrift as he dived to free a hand-line. And in some European dives, some apparently fit people have died from heart attacks at crucial moments in a dive.

Lessons are to be learnt from all kinds of accidents and tragedies. The most fundamental one is 'Know yourself: Know your equipment'. There is only one way to do this and that is to join the Cave Diving Group of the British Speleological Association whose headquarters are at Settle. Conventional sub-aqua training is given but this is only part of the programme designed to cope with the special contingencies of cave-diving. Ordinary standards are not enough, not even the passing of the tests which have been designed. Ken Pearce stresses that passing the tests and the ability to make simple cave dives does not necessarily make a competent cave-diver. He envisages situations where it might be necessary to remouth and replace mouthpieces during a cave-dive or, under some circumstances, to share a lung. As he points out, these are simple enough operations in open water, but entirely different problems in the often cramped and complex conditions of a cave. Having stressed the risks involved, I shall here confine myself to sketching the bare outlines of the requirements for cave-diving and urge anyone keen enough to seek further advice and training from the Cave Diving Group.

There are many self-supplying diving kits in use today, many of them designed for the sea. Cylinders vary in size and contents. But the Cave Diving Group itself recommends the re-breathing system using oxygen from high-pressure cylinders with the carbon dioxide waste product being absorbed by soda lime. It is recommended that water depths of thirty feet should not be exceeded with this set. Oxygen cylinders are less bulky than compressed air, which is another benefit when working in the confines of a cave passage.

Cave-diving training involves a complete familiarisation with all parts of the equipment to be used. Dismantling and reassembling are performed until the whole operation becomes automatic. When you are stuck in an emergency is no time to be wishing you had done a little more training. It is important to know the physiological problems of diving and how to meet the various physical difficulties involved. It is not likely that a British cave-diver will be affected by the 'bends' or caisson disease, which is caused by pressure in the blood being reduced too rapidly leading to its nitrogen content trying to escape in the form of bubbles. Few, if any, caving clubs would ever have the need for the appropriate decompression unit to treat this painful and often fatal condition.

Cave-diving equipment is obviously expensive, far more so than normal caving gear. Apart from the cylinders and harness, a wet suit must be worn. Sometimes progress can be made by walking under the water. At other times, free swimming is the best way of progress for which flippers are needed. Other equipment for further exploration by conventional methods must be carried through for, as Dr Pearce has said, 'Flippers are no use on ladders.'

More so than in ordinary caving is it necessary to find one's way and to keep close checks on depth and direction. This is

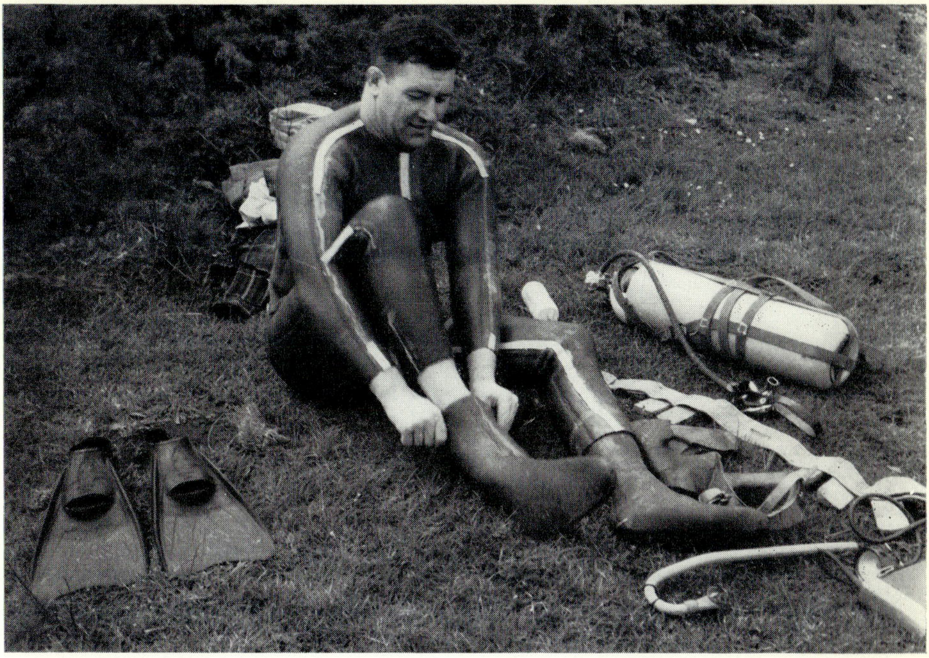

17 *(above)* Cave-diver Dr Ken Pearce preparing to explore the flooded entrance to a Yorkshire cave and (18, *below left*) about to make the descent. 19 *(below right)* A potholer coming up for air after diving through a 'duck'. Good lung power and steady nerves are essential; panic and you are in trouble

20 A telephone control point during a cave rescue

21 How not to end a caving trip: a young caver, shocked and suffering from exposure, is firmly fixed to a stretcher

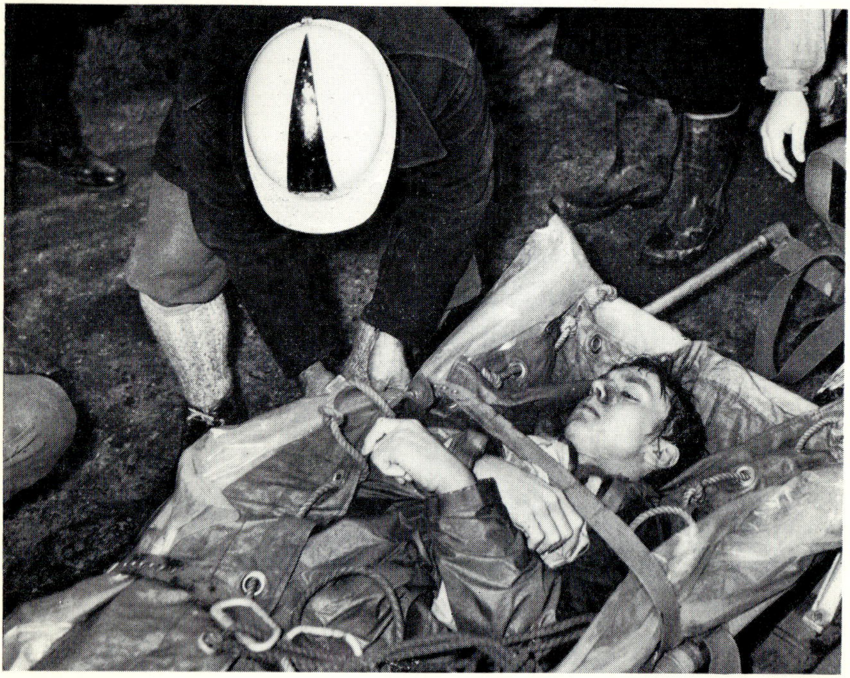

22 An instance of extension flash in a cave to give greater depth to the picture. Taken during a night exploration of Peak Cavern, Castleton, Derbyshire

23 *(above)* The author during an exploration of the inner reaches of Peak Cavern. 24 *(below left)* Stay-down expert and specialist in survival Geoff Workman in the entrance to the Mud Chamber in Gaping Gill, Yorkshire: 25 *(below right)* Miners' implements and tubs, relics of the time when they worked for eightpence a day in the Blue John Mine, Castleton

usually done by means of an all-purpose gadget which solves these problems, known as an 'aflo'. Basically it is a buoyant platform on which are placed the instruments which will give the necessary readings of depths and direction, etc. The accepted design is that which includes a powerful light, a depth gauge with the critically dangerous depths clearly marked, a compass (preferably one which does not stick when tilted away from the horizontal) and a waterproof watch. Some divers prefer to carry these things fastened to their wrists. 'Aflos' also carry a drum or reel of calibrated line or wire to indicate the distance run out.

If individual lighting is taken, not fixed to an 'aflo', it must be remembered that some of the accumulator-type lighting sets have filler holes in the sides through which water can seep and affect the battery. These holes must be plugged with plasticine to keep the water out. Other recommended personal equipment includes a sharp, stainless-steel knife, wire cutters and spare torch.

Lifelining technique must be learnt and mastered but the greatest mastery necessary is over oneself. The worse enemy of the cave-diver is panic. This arises from fear. The only way to overcome it is to train as if your life depends on it—as indeed it does. As Norbert Casteret so rightly said: 'One's life hangs upon a thread in those narrow tubes of rock filled, or almost filled, with water.'

8 *Food and survival*

It cannot be too often stressed that caving should never be undertaken light-heartedly. I am not being portentous when I say this. Cavers generally have a well-developed sense of humour as cartoons and writings in the caving journals reveal. Nor am I suggesting that caving cannot be undertaken without a sense of fun as well as a spirit of adventure. But what I am getting at is that every descent must be made with an awareness of the risks involved, not only to oneself but to others in the party and also to possible rescue teams as well. It is important to know the basic means of survival in the event of an accident or the cutting-off of an escape route to the surface due to flooding, roof-fall or, as has happened, a ladder being taken up due to bad leadership in not checking how many of the party have gone down and how many have come up.

Cavers lavish care on their equipment but tend to overlook what is often a most important aspect of caving—nutrition and survival. For very often a flash flood or a sudden thunderstorm can turn the most innocuous excursion into one of survival. I have seen waters rise above and below ground at phenomenal rates and this has led to many disasters. When one is presented with this kind of emergency, food and liquid are important factors in survival, first to keep up one's strength and energy reducing one's own liability if a rescue is necessary; and also to keep up morale. No one deterioriates more quickly under conditions of stress than a cold, hungry and thirsty person.

Some cavers limit their underground nourishment to a few jam sandwiches and a flask of soup with little thought to the nutritional needs of an active caver or the potential deterioration due to an enforced and prolonged stay underground. Few remember to take extra water in a clean container with them. The water requirements of people vary, but the average normal daily requirement of a caver is between two and five pints. Cavers lose a lot of body liquid by sweating due to the heavy clothing worn and the tight rubber suits or plastic suits in favour these days. Other loss comes from the passage of urine and the exhalations of the breath. It may sound ridiculous to suggest taking water along on a caving trip, especially as so much abounds underground. But people have been cut off in dry areas of caves without even the drips from stalactites to afford relief. It is a useful tip to have a short length of rubber tubing which can reach at otherwise inaccessible sources of supply—but it is also worth bearing in mind that underground streams can be polluted from the surface inflow. It is also a good tip, on larger expeditions, to include a ration of salt in tablet form allowing a ten-grain tablet in each pint because if the body is short of salt it cannot retain water. There is a tremendous loss of salt in sweating and it must be replaced. It is also advisable to take the salt in solution in water rather than in its solid form as this can cause gastric and intestinal troubles. Dehydration can play an important part in the onset of fatigue becoming crucial over a long period without liquid so that it is a good idea to take at least a third of a pint at two-hour intervals, preferably sweetened with sugar or glucose, to ward off fatigue. Even if you do not feel thirsty take a drink every two hours because a feeling of thirst is not an entirely reliable indication of dehydration—the process can start without the onset of a thirst.

The food taken should not be just thrown together haphazardly but should be balanced to meet the requirements of the immense output of energy underground and also the needs of an emergency. Mr Arthur B. E. de Jong, the Development Physicist of

Horlicks Limited, whose products I have used on expeditions above and below ground, and who has so kindly given me the benefit of his immense experience, suggests that a target figure of a balanced underground daily diet of 4,000 kilo-calories, should be aimed at. For prolonged strenuous work like laddering, crawling, digging, etc., 5,000 kilo-calories. This should be correctly proportioned as between foodstuffs containing protein, fat and carbohydrates, in addition to the minerals and vitamins. Proteins are essential for the muscle and tissue replacements, fats for energy and heat and carbohydrates to provide the balance. On shorter expeditions minerals and vitamins are not so important but on the longer stays underground they are vital. Among other things they help to combat skin disorders which occur due to a lack of sunlight, poor diet and other disabilities which occur due to the lack of opportunity for personal hygiene. Food gets used up rapidly in the body's engine on a caving trip.

Breakfast is probably the most important meal on which to start the caving day. In fact it is vital to have the inner man well fuelled before making a descent. It is not unknown for some cavers to do a fair amount of drinking on the night prior to descent and the consequent hangover affects the appetite. On anything like a serious cave expedition, it is wise to leave liquor alone for at least twenty-four hours before setting out underground. Alcohol, apart from its other vitiating effects, contributes greatly to dehydration and the onset of exposure. I know of one death from exposure in which a heavy drinking session the previous night and no breakfast the following day were the main contributory factors in the onset of a fatal coma. The person concerned just literally laid down and died because he had no reserves of energy from within to combat the rapid onset of physical deterioration. An average-sized breakfast would probably have saved his life.

Cavers invariably have healthy appetities and the breakfast which has been the start of all my climbing and caving days—porridge, bacon and eggs, beans and fried bread plus liberal lashings of tea—contains enough calories to produce a peak energy level at one and a half to two hours, from when the energy curve begins gradually to decline. Replenishing should take place, if possible, in up to two to three hours after the breakfast (or supper of equal value if it is a night descent); if the trip is to be twelve hours or over, arrangements should be made for at least one substantial cooked meal at some suitable rendezvous. Choose food which gives a good power-to-weight ratio working towards, say, 11-14 per cent protein, such as meat, eggs, fish, milk; 25-35 per cent fat (meat, butter, margarine, lard or cooking lard); and the rest in carbohydrates such as flour, cereals, bread, sugar, etc. Leaders should also insist on each member of a party carrying a small personal emergency pack which, wisely used, can sustain a reasonable degree of survival for a forty-eight hour period. Such a pack, containing two and a half ounces of oatmeal block, one and a half ounces of glucose tablets, and two ounces of barley sugar, carefully rationed with two pints of water has been devised and tested and will do the trick.

Alternatively, if the party is large and sharing loads for an extended stay, Horlicks' one-man, two-day rations should be carried in sufficient quantity for the party. Such a ration packs into a size of ten by four by two inches, and has a total weight of approximately three pounds. The total ration supplies 5,167 kilo-calories. Used efficiently this ration is adequate for two days and may be extended to five days should the emergency be so grave that immediate rescue appears doubtful—provided that water is available, and once again water is stressed as a vital factor in this aspect of survival. These survival packs contain one oatmeal block, two packets of Horlicks tablets, two packets of enerzade glucose tablets, one block of fudge, rum-flavoured, three milk chocolate bars, four expedition biscuits, one block of processed Cheddar cheese, one ration of meat-and-vegetable bar with potato powder, a wooden spatula, a polythene water-cum-food bag and one packet of booklet matches. This ration has been devised so that all the food items can be eaten uncooked. An active party could take an oatmeal block for breakfast, biscuits and cheese for a snack lunch, chocolate for a main meal and during active operations the fudge. The oatmeal block can be eaten uncooked. It could be made into a porridge by crumbling half of the block into about a quarter of a pint of water, bringing to the boil until it becomes a normal thick, smooth porridge. The Horlicks tablets, which can be sucked slowly while resting or marching, help to alleviate hunger, and the enerzade glucose tablets, also used on the march, are one of the best foods for reducing water requirements during long dry passages underground. Again the fudge can be eaten dry or made into a drink by dissolving in hot water. Apart from being used as a snack meal the biscuits and cheese can be cooked: a biscuit can be soaked in water and fried with a layer of cheese on top. The meat and vegetable pack with the potato powder will be sufficient for two main-meal servings. The potato powder, apart from being reconstituted into creamed potato by adding a small quantity of boiling water and stirring, may be used as a thickener for the meat bar. The spatula of course can be used for cooking purposes or for dividing food into suitable portions. The water-cum-food bag is useful for carrying unconsumed food. In addition the wrappings of the food may also be used as a fire-lighter if such heat is needed. I have full confidence in this type of ration because experiments have been carried out extremely thoroughly with particular consideration to its application to caving. The food will store well. From 1961 extended storage tests were carried out in Agen Allwedd in South Wales. This is one of the largest cave systems in Britain which is liable to flooding and parties being cut off. Tests showed that these rations were able to survive in this cave up to three, possibly four years and still be satisfactory, having been placed in caches in various parts of the cave in the event of an emergency.

One of the most frequent emergencies in caving arises from being cut off by water or the careless leaving behind of one or two members of a party. To survive needs concentration, common sense and the control of any feelings of panic. Being trapped by water really should not happen—but it does. Caves which are known to flood should be avoided at the slightest risk of heavy rain. No one knows in this country when a simple rainstorm will develop into something of monsoon proportions. When it does happen the effect above and below ground is awe-inspiring.

Once, during a dig at an old mine at Snelslow in Derbyshire, one such storm swept the Peak District; if I had not got a series of photographs, some taken neck-deep in water, I still could not believe it really happened. A tiny stream runs into Snelslow which, I suspect, breaks into the Giant's Hole System and with members of my own potholing club we had built a small concrete dam at the entrance to divert the stream water into a secondary passage. We also had a sign 'Do not enter when flooded' thinking that the water rose only just above the sill of the dam. What we had not bargained for was that not only did the dam disappear under water—the complete cave entrance was submerged by a raging torrent which filled the valley well up the side of Snelslow and disappeared in a great vortex down a swallow hole beyond. It was a terrifying example of the water power which can isolate cavers and lead to such tragedies as Mossdale Cavern in Yorkshire in 1967 and the disappearance of a lone caver in Goyden Pot, Nidderdale in 1957—the twenty-year-old Leeds art student, Brian Andrew Kerr.

There are a number of useful points to bear in mind if trapped in a cave for any reason. One may be alone or with a small group. The first and vital essential is to keep calm and evaluate the situation. If water is rising it is important to get to a place beyond the reach of the rising torrent. If one has just been thoughtlessly left behind, it is best to get to the nearest possible point of rescue. With proper leadership no one *should* be left. There was a case in Derbyshire recently where a caver was completely overlooked and it was some hours before he could be retrieved. Flooding creates a different situation and it could be days before a rescue team can successfully get in, depending on the weather and the strength of the rescue resources. Talking to Lewis Railton of the South Wales Caving Club of the occasion when he and Bill Little were cut off for fifty-nine hours in Ogof Ffynnon Ddu in 1951, I asked him for some of his impressions. Significantly one of the things he stressed as most important was the need to carry a reliable, waterproof watch, preferably one with a date window. The reason for this, he told me, was that without a watch there is no means of measuring time and when it comes to rationing food out this is a crucial factor. An hour seems like an age and there is a temptation to eat food at the dictates of one's stomach rather than to spread the food out at very definite intervals. As he pointed out, 'without means of recording the passage of time three days' ration of food could be eaten in twelve hours.' If no watch is available it is a good idea to revert to one of the oldest forms of measuring time—marked candles. All cavers should carry spare candles, which have a fairly standard rate of burning providing draughts can be excluded. It should be well worth while at some Club meeting to establish on a candle the rate of burning of say an hour and mark it. Then it would not be difficult to produce a number of graduated candles with snicks or black lines to give a fair idea of how time is passing.

Having reached a dry point of the cave, get off what damp clothing you can and replace it with any dry gear you have. If a towel is carried—there should be one at least in a camera kit—have a brisk rub down before putting on the dry gear. Again, it is important to conserve energy, so while the emergency lasts it is advisable to sit or lie down so that little as possible of the energy-producing food and drink is consumed in wasteful physical action. Try and get warm and stay warm. If there is a draught, try and build a small shelter with stones and mud but, particularly if the cave is a confined space, do not dig a trench to lie down in, especially in sand, as this can cause a pool of exhaled breath in the form of CO_2, carbon dioxide, to build up. This in itself can start a vicious circle by speeding up the respiratory rate and adding to the physical and mental deterioration with headaches, general weakness and dizziness. It does not require any great diminution of oxygen to affect the nervous system. You are in trouble when the oxygen from the inspired air drops from the normal 21 per cent to between 16 and 12 per cent; at that moment anoxemia develops—oxygen deficiency. In such a confined space a candle or any other naked flame is a bad thing—but it can give a guide to the danger of the situation as it usually burns out at 17 per cent. Deterioration comes on in many ways. One of them is the psychological lowering of morale due to exposure and the environment. Cave malaise or sickness can be a very real thing. I would also recommend trying to find a quiet part of the cave system in which to sweat it out away from the roar and rush of water. As times goes on the unending drumming of water can get on one's nerves. It seems to numb the senses and the knowledge that it is rushing along to the outside world increases the temptation to jump in and go along with it to try and find a way out that way. Watching it continually has a hypnotic effect—although a regular watch should be kept as to whether it is rising or falling. Professor Alfred Beögli, who was trapped for ten days in one of the largest cave systems in the world, the Höloch in Switzerland, with four men said that immense mental discipline was necessary to refrain from trying to get through the flood waters. His party had only gone in for a day, unprepared for a longer stay, and their escape is a reminder of what can be achieved through not yielding to panic and eking out food supplies, however meagre.

If you do move into a deeper and drier part of the cave, leave some indication of where you have gone for the rescuers who come along. White cards in polythene covers are useful to leave around in some eye-catching spot, or an arrow made with stones, or as a last resort a message scorched on any dry part of the cave wall with a candle or carbide flame. All these of course presuppose that there is no telephone link to the outside of the cave and no stand-by surface party.

With a party there is always a temptation to separate, particularly as tempers do not improve as the strain builds up. Decide on a leader to cope with the emergency and stand by his judgement. And keep a log if you can of all the things that go on. Write down the times of food taken, when the next meal is planned; also any physical changes in the atmosphere of the cave, the time water levels rise and fall, conditions of personnel. These are disciplined acts which absorb time and help to concentrate the mind away from the situation. And if there is no danger of a CO_2 build-up, sing. One of the greatest dangers to a trapped party is psychological, and singing or chanting anything from the Rubaiyat of Omar Khayyám to Eskimo Nell can do no end of good. One thing is certain: no matter how bad things seem, a rescue team will be on its way. You owe it to them to keep as fit as possible with a good morale for when the moment comes to make the journey back to the surface. In that way you can help yourselves and help them.

9 *Cave photography*

Photography is one of the most fascinating hobbies to be associated with caving. There is a wealth of colour underground as well as wonderful opportunities for studies in black and white. But cave photography poses problems of a special nature. Most of the enemies to good photography exist underground—damp, dirt, dust and steam. There is also the inevitable banging, shaking and immersing in water the equipment must undergo on any trip underground. To cope with all the hazards to which it must be subject, the equipment—cameras, lenses, flash guns and all the other sensitive apparatus of photography—must be robust. The equipment must also have the maximum protection in order to survive and do its work.

The main emphasis in this chapter will be on still photography, as cine photography underground is expensive and unpredictable. There have been some excellent cine films made but the equipment and the film are usually beyond the financial reach of the average caver. Obviously, the first essential of photography is the camera and, as there is so wide a range from which to choose, I can only write in general terms based on the experience of myself and my friends. The two most convenient sizes of camera for cave photography are those which take films giving a format of $2\frac{1}{4}$ x $2\frac{1}{4}$ inches (6 x 6 centimetres) which are 120 or 620 spool size; and the 35-mm. cameras of which there are so many types and makes. My personal preference is for the $2\frac{1}{4}$-inch square format for black and white pictures and the 35-mm. size for colour transparencies.

The 35-mm. camera has the advantage of producing a large number of pictures fairly cheaply. Quite a number of 35-mm. cameras are also inexpensive and some of the less costly variety offer the facility of interchangeable lenses—wide angles, telephoto and close-up—which are extremely useful underground as they add much versatility to one camera. For many years I have used an old, battered Solida III with a 2.9 lens and a Prontor S shutter with a self-taking device, which is very useful for getting into one's own pictures—if you do not fall over or trip over a boulder when rushing to get into position. This camera has been with me to the Himalayas and has survived being blown away in a whirlwind, complete baptism in Giant's Hole and a fall of sixty feet in Nettle Pot. It is a folding camera, which some cave photographers do not care for, but it fits into a fairly small space and travels well. I have experimented at various times with twin-lens reflex cameras but, generally, these are too bulky for cave photography although they do make focusing and composing easier. They also have the advantage of taking interchangeable lenses, and the automatic wind-on eliminates the fiddling business of looking through a tiny rear window, often in poor lighting, to see if the film has been wound on to the next frame. Nearly all 35-mm. cameras have the automatic wind-on which also sets the shutter.

The ideal caving camera has yet to be devised. Shutter speed (unless electronic flash is used) is not important because most pictures taken underground are done on Time (or open flash) or on the Bulb setting. Depending on the type of flash bulb used, the most usual shutter speeds are of the order of one-eighth to one-thirtieth of a second. Higher shutter speeds are useful to 'freeze' waterfalls to give better water definition instead of a streaky blur, the stopping of any movement such as someone abseiling, and to eliminate camera shake. But as it is wise to take most pictures from a tripod, this last problem should not arise. Having mentioned bulbs and flash guns, it is important to have a flash-gun plug point on the camera, although pictures can be taken without having the gun on the camera.

Another factor in the choice of a camera is the size of the viewfinder. Some cameras have a not very adequate little hole through which to peer. Unless the scene can be brilliantly lit beforehand for it all to be seen through the viewfinder, it is difficult to compose a picture through such a hole. To overcome this, some cameras do have direct or sports viewfinders. They are not difficult to make if the camera is without one. For this reason I have recently switched to one of the new single-lens reflex cameras, the Praktisix, which gives an excellent view and also takes interchangeable lenses and is synchronised for electronic and bulb flash. It also has the rapid-wind lever found on the 35-mm. cameras which sets the shutter and advances the film one frame. It is, however, somewhat heavy and also quite costly. But it is a luxury in which I feel entitled to indulge having started my pictorial career underground with an ancient Thornton-Pickard stereoscopic camera superbly made in mahogany and brass, for which I had double-plate holders dated 1888. Yet its rapid rectilinear lens produced some excellent pictures with 3D effects. Alas, its piano-accordion-type bellows were unmendably punctured by a piton and it passed into my caving club's museum.

Of the 35-mm. cameras, I find it impossible to recommend any particular one. Photography these days is surrounded with a curious mystique and there are untold numbers of books about this and that aspect of it. But I have found that simplicity in equipment is a great asset. Unless one is a specialist in optics and is hypercritical to the *n*th degree, it is difficult to detect the difference in performance between the lens in the cheaper 35-mm. cameras and those which cost well over the £100 mark. Admittedly, the cheaper cameras have only fixed-lens mounts (although these are quite adequate for beginners). In the lower price range I have found the Exa cameras are becoming popular with many cavers because it is possible to build up a range of accessories which can cope with any number of problems found underground. In the past two years I have extended my equipment from the standard 2.8 Tessar 50-mm. lens by adding a 3.5,

30-mm., wide-angle Meyer-Gorlitz (bought secondhand for £5) and one of the same firm's 2.8, 165-mm. tele lenses. Probably the most useful of these lenses is the wide-angle, which, used in vertical pictures, can give those extra feet of height in addition to the width, where needed. Wide-angle lenses also have great depth of field because of their short focal length, which is so useful in cave photography, and they can also get in extremely close without the need for a close-up lens. The angle of view varies according to the type chosen but the average is between 45 degrees and 60 degrees and it can be as high as 93 degrees.

If you cannot afford a camera which takes the different lens, have a look round for a good secondhand Ilford Advocate, which has been popular with many cave photographers. I understand that the Advocate has been discontinued, but there are still plenty around. A friend of mine swears by his Advocate and has consistently obtained first-class results with it. It has a 35-mm. focal length which gives a fairly wide angle of view and an excellent depth of field.

Another useful asset on a camera which is not of the reflex type is the coupled, split-image range-finder. This is a big help in focusing on, say, a stalactite, headlight, candle or any other distinctive object. While on the subject of focusing, do not forget to pack a tape measure in your kit. When all else fails, this can be most useful in establishing the correct distance between camera and subject.

Having got the camera, you will need something on which to stand it. With some pictures, working at 1/25th of a second or at faster speeds with electronic flash, a tripod is not necessary, although people can get camera shake at 1/25th if they have been doing some energetic caving immediately before taking a picture. But for nearly all other exposures, except where it is possible simply to stand the camera on a convenient piece of rock, a tripod is necessary. Apart from helping to eliminate camera shake, it is a help in composing pictures and vital for exposures where more than one flash is needed. To withstand the battering it will get underground, a tripod must be sturdy. Weight is important. A few extra ounces feel like pounds at the end of a hard day but I think it is worth the slight inconvenience of weight in order to get the extra stability. My longest-serving tripod is too heavy for some, but it has withstood being dropped past my head in the Gouffre Berger and is still going strong. Remember always to clean your tripod, and indeed all your equipment, after each trip. Mud comes off so much more easily if it is not given time to set solid and immobilise the legs. Sluicing down with water at the first opportunity is the best way of getting rid of the mud. A light application of petroleum jelly afterwards helps the moving parts enormously. Some tripods also have tilt and angling heads—otherwise you can buy a ball-and-socket head which gives these movements. Another useful aspect of the tripod is that you can use more than one camera at once, thus economising on flash bulbs by using one flash for two or more cameras fixed to a bar on the tripod head. In this way you can take black and white and colour shots of the same scene, obviating the expensive process which is necessary to make black and white pictures from colour transparencies. This can be done by using two films of approximately the same speed rating or compensating for any difference in the speed by an adjustment of the stop. A tripod with a bar can also be used to have the flash gun a little way from the camera to avoid some of the flat, frontal lighting one gets from a flash gun directly on the camera.

Flash guns again come in so many shapes and sizes that it is very largely a matter of personal choice and preference in deciding which gun to use. The most expensive is the electronic flash gun which discharges a gas-filled tube giving a bright light of short duration. Some use chargeable batteries, which more or less precludes their use underground because of the re-charging difficulties. There are others with replaceable dry batteries and they give anything from sixty to about 100 flashes per charge. On the face of it, this is cheap lighting—but the guns themselves are expensive, and there can be a danger in using the high voltage they use in caves, where the humidity is usually high. Electronic flash guns are also heavy in comparison with those using bulbs and there is a risk of breakage. Lighter, more compact and therefore more suitable for caving are the flash guns using bulbs. The current used to fire a flash bulb is not high and the best ones use a battery capacitator circuit using a twenty-two and a half volt hearing-aid battery (B 122 in the Ever Ready range). Quite simply, the capacitator is charged by the battery sending a current through the bulb with the close of the synchronising contacts on the camera. For caving purposes, look for a gun which has a detachable reflector like the Zeiss Ikoblitz, which fits snugly into a plastic case. Others like the Kobold U 1 fold up fan-like and take up very little space. Nearly all guns of this type have an ejector device which saves you getting burnt fingers when removing the bulb. It should also have a built-in test bulb which lights up to indicate that the circuit is working satisfactorily. Another useful type for open-flash photography also has a button attachment for firing independently of the camera release. This allows for a number of flashes, useful in a big cavern or in a long passage. Finally, there are the big guns, now getting out of date, which use the large, screw-in type bulbs. They have much bigger reflectors and give an immense amount of light to cover a large area.

The bulbs most in use underground are PF 1s or PF 5s and are capless. Some flash guns will only take this type, but there are others which take the bayonet type and there is an easily fixed adaptor which will enable them to take the capless bulbs as well. Incidentally, a quick lick to moisten the end of a capless bulb helps enormously to clean the end of the bulb and improve the contact. Flash cubes were brought in about 1966 with their own built-in reflectors but they are useful only for close-at-hand subjects. Blue flash bulbs are the best for all-round work. They can be used with daylight colour film, the best all-round film which can be used for shots outside the cave and inside it as well.

The other sources of illumination are flash powder, magnesium ribbon and available light from candles, torches and headlamps. Flash powder is now passing out of vogue. Some cavers swear by it. But it is an unstable substance. Manufacturers may deny this but I have had some unhappy experiences with it. I lost my eyebrows and a lot of my once curly hair in making too close an examination of a mound of powder which had not gone off when the fuse reached it. Another disadvantage is that in spite of the description ' smokeless ', flash powder does tend to fill a cave with ' fog ', which percolates along the various passages, making further photography impossible until it clears. If another party is in the cave hoping also to take pictures, this can provoke a certain amount of bitterness and unpleasantness. But flash powder has its uses in being able to illuminate huge areas which flash bulbs may not be able to reach such as the main chamber at the bottom of Gaping Gill. It is important to keep the flash powder dry and away from naked lights; always be extremely wary of a misfire. There is less chance of precision in exposure with flash powder and I have never been happy with it. Apart from the smoke involved, the explosive nature of the powder has been known to damage fragile formations and, in some unstable cave conditions, it could produce an unpleasant rock fall. Magnesium ribbon also has the disadvantage of creating smoke which hinders further photography and the exposure is usually of the by-guess-and-by-God nature. The light it gives is attractive and can show large areas of cave but it is of a brilliant nature and it takes time to reaccommodate the eyesight to the normal pothole lighting from headlamps etc.

Makeshift lighting is used only as a last resort when all other lighting sources have packed in. It is possible to take pictures, undoubtedly of long exposure, by the light of a headlamp or candles. Again, this involves a fair amount of guesswork.

Cameras, tripod, flash equipment—these are the basics. In addition you need a release cable for the camera shutter, preferably one with a locking device with which to keep the shutter open. Other useful gadgets are extension cables, which vary in length and give mobility to the flash unit. It is often better to move the flash gun away from the camera to get better modelling into the pictures as frontal lighting tends always to be flat and hammers out a lot of the detail. Another gadget which helps when more light is needed is the triple-headed contact which fixes to the camera flash point and allows three guns to be fired off at once, when necessary.

A lens hood is a 'must'. I use one with a plain, optically correct glass in a filter mount to provide extra protection for the more delicate surface of the lens itself. It is much easier to clean this than the lens as it helps to keep out mud, dust and damp. A soft brush and lens-cleaning tissues should also be included in the kit, because in cave photography clean equipment is of paramount importance.

Some people may think it is 'cissy' but I think it is a good idea for cave photographers to wear rubber gloves most of the time when caving. They can be worn right up to the moment of setting the camera, leaving the hands clean for the necessary final, delicate adjustments. Ordinary domestic gloves usually suffice, or thin, surgical gloves can be donned to keep dirt away from the equipment at the requisite moments.

Protecting the equipment is another problem. Ex-Army ammunition boxes make excellent carrying cases and there are various forms of plastic material such as foam rubber which can be used to pack cameras, lenses, etc., securely in position. The custom-built carrying cases ideal for above-ground photography would not last the strain of hard, underground transportation. Polythene bags are ideal for keeping moisture out although they do have a tendency to 'sweat', thus creating condensation problems. One way of combating this is to use Silica gel crystals which absorb some of the moisture. The best form is with an indicator which gives a warning that the gel has become saturated and is no longer doing its work. While on the subject of condensation, it is one of the hardest things to combat in cameras used in caves. It can occur on the lens and on the film when cameras are taken from colder to warmer atmospheres. Usually caves, certainly during most of the year, are cooler than the outside temperature. But condensation can also come from exhalation of breath and the steam which emanates from a caver's wet clothing. Consequently, it is as well to limit the number of people in a photographic session, especially in a small cave, where many bodies can create an impenetrable fug. If a fairly big party is involved, it is wise to get the pictures first before steam can develop, otherwise pictures can look like astral phenomena or a London fog. Always check your lenses, avoiding at all costs breathing directly on to them. It is no use wiping off condensation when it is forming just as quickly again, and while you may be able to get at the front it is often impossible to wipe the rear of the lens with the film in position. One can only wait patiently for this to disappear although some people do try to speed up the process by the judicious application of the flame from a carbide lamp. This may be all right in certain desperate circumstances, but the heat treatment can be damaging to a lens and the camera body. And when all the moisture has been cleared some pictures have been ruined in the last split second of taking by the photographer's own breath, particularly when firing from eye-level. This is where the tripod and a long release cable are of such value. If you are firing from eye level, hold your breath at the moment of releasing the shutter. Also check your lens frequently for other forms of moisture—droplets from above or spray from waterfalls or fast-flowing water. This kind of water can be removed with a chamois leather if you wish, but remember that a lot of cave water contains minute grains of deposit which can scratch a lens surface. That is why a plain optical-glass covering is so useful.

Photographers can argue for hours on the choice of film. Some prefer soft results in their black and whites, others like severe contrasts and high-quality sharpness. The various colour films also have merits and de-merits. It is best to experiment a little and then stick to one type of film until you know its characteristics under all kinds of conditions, its latitude in terms of exposure and also its performance under various development techniques. For this reason, I leave the question of choice of film wide open except to say that I have worked with Kodachrome and High Speed Ektachrome, Kodak Tri-X and Ilford HP3 with good all-round results.

Because so many things can go wrong, cave photographers must accept disappointments and failures. Frequent ' inquests ' are useful to find the causes of failures and to eliminate them. A number of factors are involved in taking a shot—the speed of the film, the power of the flash involved, the distance between camera and subject and the stop to be used to attain the desired result. The stop, of course, also involves the depth of focus, and you must decide how much of the scene you are photographing is to be sharp and set your stop and distance accordingly. The manufacturers take some of the labour out of the calculations involved by producing their flash bulbs in packets printed with guide numbers with tables of apertures relative to distance and film speed. To get the aperture you want get the guide number and divide it by the distance between the flash and the subject—do not forget it is the distance of the bulb from the subject and not the camera that matters, especially when working flash off an extension. This is where the tape measure comes in handy. For instance a PF 5 B blue flash blub with a film of 64 ASA filming at ten feet from the subject at 1/25th to 1/30th of a second would work out like this —Guide No. 170 divided by distance of ten feet equals seventeen. As the nearest stop to seventeen is sixteen, the picture would be taken at f16. Remember, it is always better to overexpose than underexpose and when the aperture worked out does not coincide exactly with the stops on your camera, take the aperture below the one worked out rather than the one above. Eventually, you will come to work out your own standards based on experience with a particular film and flash bulb. All exposures based on manufacturers' guide numbers are geared to average room-type conditions and take no account of the varying conditions underground. Some of the white formations underground add greatly to the light reflection and can make a great difference to either the speed or the aperture used. The glistening effect of water also increases the light reading. In fact some direct flash light on a subject is too much, causing loss of definition, and produces ' muddy ' pictures. Newspapermen have developed a technique of ' bounce ' flash, directing the flash to the ceiling or wall so that less light falls on the subject. Some cavers now use the Paraflash method, which is quite pleasing in smaller caves. This involves the use of a white nylon parasol at which the flash is fired. This bounces back a more diffuse lighting and, if using one of the proprietary models on a tripod, the distance between the gun and the parasol is constant and it is not difficult to work out the different exposures or apertures after a few experimental shots.

In cases where two or more bulbs are being used at roughly the same distance from the subject but from different positions, the guide number should be increased by fifty per cent for each bulb added. But if two flash bulbs are used from the same point, the guide number should be doubled. Using more than one flash bulb involves extension wires or the flash gun which can be fired independently of the camera. With the independent method it is best to have other cavers, along a cave passage or concealed behind rocks, point the flash guns in the required direction. Then the shutter is opened and the various flashes are fired on a word of command. It is best to practise this beforehand for there is nothing more likely to aggravate a cave photographer than to have the flashes fired before his shutter has been opened. Multiple flashes fired at the same time can cover a huge area and these can be fired from a distribution box either home-made or bought. Having worked out your exposures, remember also that there is a square law of fall-off from the light source. F16 at five feet will become f8 at ten feet and f4 at twenty feet.

These, then, are some of the broad outlines of cave photography. The subject itself would fill a book and then can only be a general guide. It is wise to keep a log-book of all shots taken, place, time, film, exposure, conditions, etc. From this kind of data you can see where you are going wrong or how your successes are achieved. There are many things that can go wrong, and in the end, as with so much of caving experience, you will only learn by trial and error.

10 *Where to cave*

Non-cavers often say to me 'What do you see in potholing?' To reply fully would take a lecture at least. But apart from the interest of underground exploration, I also make the point that cave exploration has taken me into some of the loveliest country in Britain and in many other lands. All the limestone regions of the world have a special beauty of their own and cavers can enjoy the best of two worlds—above and below ground. Throughout the world there are literally thousands of caves and potholes, many of them still awaiting discovery or exploration. And this is part of the fascination of speleology. In Britain, where all our mountain summits have been climbed, every inch of the land surveyed and chartered, it is still possible, underground, to enjoy the thrill of standing where no human foot has ever trod before.

Europe, Asia, the Americas, Africa and Australia—and, I believe, the Soviet Union—have unending possibilities for exploration. In the Himalayas I have found what I think is the world's highest known cave in the limestone which forms part of the Everest *massif*. High on the side of Nuptse, one of the trinity of Everest peaks, there is a cave at about 24,000 feet. It could be the ultimate caving problem as it would involve a major expedition just to get there, let alone get into it.

The main cave regions of the British Isles are in Derbyshire, Yorkshire, the Mendips, South Wales, the Furness Coast and Devon. There are a few caves in Scotland and a great number in Ireland.

DERBYSHIRE

The Peak District is right on the doorstep for hundreds of potholers from its neighbouring towns and cities. Every variety and classification of cave and pothole can be found within a few miles' radius of the hub of this National Park at Castleton. The underworld of the Peak has been known since the earliest times and there are records that the Anglo-Saxons worked the lead ore in one of the country's oldest lead mines—Odin Mine at the foot of Mam Tor. Tradition has it that it was once used as a penal settlement by ' convicts '. And it was the lead miners of old who broke into the first natural caverns of the County in pursuit of ' t' Owd Man '—the local name for the Devil, who was supposed to have left the metal behind. The lead miners were the forerunners of modern cavers, achieving incredible results with their simple tackle. One of the oldest known natural caverns in Derbyshire is the famous *Peak Cavern,* which was depicted, somewhat rudely in the earthy language of those times, as ' The Devil's Arse i' th' Peak ' and was referred to as such in song and legend. *Peak Cavern* is a show cave today but its inner recesses can be explored beyond the tourist limits with permission from the owners, the estate of the Duchy of Lancaster. Part of the system beyond Pickering's Passage is sealed off for all time, for it was in this section that Neil Moss became entombed. *Peak Cavern* is still interesting and has been the scene of much hard work by cave-divers who have helped to solve most of the cave's secrets.

Castleton is famed for its show caves, which give a comfortable introduction to caving to tourists without any of the inconvenience or hazards. There is *Speedwell Mine,* which now has electrically powered boats for tourists to travel to the so-called Bottomless Pit; *Treak Cliff Cavern,* which was the first cave in Europe to be lit by electricity; and the *Blue John Mine,* where the rare fluorspar or Blue John stone with its delicate colouring was found and worked by the Romans.

Bagshawe Cavern near Bradwell village used to be a show cave, but came to be regarded as a little too difficult for the average tourist. It is now a useful cave in which novices can be introduced to the various potholing techniques in comparative safety. This includes a baptism in cave waters. It is necessary to obtain the permission of the owner, Mr Revell, who lives at Main Road, Bradwell; he charges a modest fee.

More sporting caves include *Giants Hole,* which is classified as a Super Severe and lives up to that classification in every inch of its extensive passages.

Some are very tight indeed, creating exceptional difficulties in the event of a cave rescue. It was in Giants Hole that I came across my only case of a psychological breakdown when a friend, whom I had believed to be a hard caver, completely lost his nerve and broke down and wept for no apparently accountable reason. Giants Hole is also one of those caves for which permission to enter must be sought. Mr Watson of Peakshill Farm, which is just along the lane towards Rushup Edge, charges 2s 6d which is not exceptional as it includes a cup of hot coffee when surfacing and facilities for changing. Giants is a very hard cave indeed and I must emphasise that it is for highly capable cavers only and not the 'nig-nogs' or 'yips' who are a general blight on the Peak District scene. This cave also links up with *Maskhill Mine*—a difficult round trip—and *Oxlow Cavern*. Permission for entry to these last two must be given by the BSA, who can be written to at their headquarters in Settle.

A pothole which is the pride and joy of the Derbyshire cavers is *Nettle Pot* which, in my caving career, has rightly been upgraded from a very difficult pothole to one in the Super Severe class. Found in a bed of nettles in 1934, it is now 524 feet deep and could go deeper. I was on the first postwar descent of Nettle Pot after its concrete lid had been blasted off. The decrepit winch was still in position above Elizabeth Shaft and, in a weak moment, I was 'persuaded' to be lowered by it. It was not a happy experience. It is still there but should not be used because it is too unsafe. Nettle has one of the tightest pothole entrances in the country where the inexperienced have become jammed, although in no great danger.

Another of the Derbyshire 'classics' is *Eldon Hole,* known as one of the Seven Wonders of the Peak. It can be laddered, abseiled and prusiked, winched or climbed on a knotted rope. Eldon figures largely in the folklore of the Peak. There is still much work to be done in Eldon. Again, this is a good pothole for training novices in big-shaft technique. As with so many other people, including the famous Dr Baker, Eldon was my first pothole and I have great affection for it—but not for the people who thoughtlessly lob stones down the shaft.

Many of the Derbyshire caves and potholes are in the area bounded by Sparrowpit, Castleton and Bradwell. Further away is *Carlswark Cave,* which is fairly strenuous and has a number of rewards for those who seek for them including splendid examples of *producta gigantea* which are a chastening reminder that this part of Britain, about equidistant between the seas on our west and east coasts, was once under the sea itself.

Derbyshire also has a number of caves of archaeological interest such as *Windy Knoll Cave* at the top of the Winnats Pass, *Elderbush Cave* in the Manifold Valley, *Thirst House Cavern* in Deepdale, Kingsterndale, which has yielded such a treasure house of Romano-British articles; and a good example of a cave-dweller's abode at *Thor's Cave*, in the Manifold Valley.

Apart from these caves, lead mines abound and some people have made a special study of lead-mining and its strange customs, which include the curious Barmote courts for regulating the industry and those taking part in it. The greatest care should be taken in all mines.

The Nature Conservancy has scheduled the following caves as sites of special interest in an attempt to preserve them and protect them from vandalism, despoilation by tipping and quarrying: Giants Hole, Eldon Hole, Nettle Pot, Peak Cavern, Treak Cliff Cavern, Plunge Hole, Pool's Cavern, the Wye Head Resurgence, and the Cresswell Caves.

YORKSHIRE

Yorkshire has been described as the cradle of British speleology. Long before the pastime became organised, the many caves and potholes of the shire of the broad acres had attracted earlier explorers, some of whom I have already mentioned. As in Derbyshire, lead miners were the first to venture into the limestone underground with a definite purpose. Even before Martel's visit, W. Boyd Dawkins and others had been probing about Yorkshire caves and potholes. John Birbeck and William Metcalfe made the first descent of what was then known as Helln Pot, better known today as *Alum Pot,* in 1847. They organised what could be called Britain's first caving team, consisting of ten people. They made the descent of what they called 'this awful chasm'—now classified as Moderate—with ropes, planks, a windlass and a fire-escape belt. It was descended again in 1848 and again in 1870. Boyd Dawkins went down with thirteen people who were lowered, two at a time, in a wooden bucket with a canopy to protect the occupants from stone fall. Also about that time Mr M. T. Farrer of Clapham, who owned the Ingleborough estates, was responsible for the earliest discoveries in what is now the *Ingleborough Show Cave*. The old cave went in for about seventy feet. Then there were great floods in Clapdale in 1837 and Farrer ordered the demolition of a stalagmite 'dam' holding back water in the cave. His servant Josiah Harrison did this in one of the first recorded acts of cave vandalism, justified, no doubt, by the discovery of more than a mile of passages and many magnificent formations. Farrer, a man of great courage, explored one of the earlier pools by swimming with a candle in his cap and a rope round his body, again one of the earliest references to swimming in cave exploration. Already, at that stage, theories and measurements had been put forward about the rate of growth of stalagmite and other calcite formations.

After the Martel descent of Gaping Gill, speleology really got started in Yorkshire, and the county now has more than 500 known and classified potholes and caves ranging from easy ones to the county's deepest *Penyghent Pot*, 527 feet deep, and the ill-fated, extremely arduous *Mossdale Caverns*. There are fourteen main caving areas in the county—Marble Steps, Birkwith, Ribblehead, Giggleswick Scar, Bruntscar, Alum Pot, Kingsdale, Chapel-le-Dale, Penyghent, Casterton Fells, Attermire, Newby Moss, Nidderdale and Gaping Gill.

Some Yorkshire caves are liable to flooding and every effort should be made to consult local conditions before going into them. The most grievous accident in British caving history happened in Mossdale in June 1967, when six experienced cavers were drowned when the system became flooded. Ten years earlier a Leeds art student disappeared during a flood in Goyden Pot, which should not be entered during bad weather or when compensation water is being released from the Angram reservoirs. The Bradford Water Authority should be consulted about the latter possibility either at their headquarters or at the reservoir itself.

Yorkshire offers the biggest cave chamber in Britain in *Gaping Gill*, and it may be entered by winch or a hard ladder climb. Like many of the Yorkshire cave systems Gaping Gill offers round-trip possibilities. One of the hardest of these circular tours is the *Providence Pot-Dow Cave* link-up which has achieved some notoriety in view of the numbers of people who have come to grief attempting it. Another is the *Lancaster Hole traverse* on Casterton Fell, a complex cave system linking up with Easegill. Another, and more recent circuit is from *Mongo Gill Hole* into *Stump Cross Caverns* for which permission is needed from both cave owners. Complicated, too, is the big and still growing system of *Lost Johns* on Leck Fell, which has a number of problems and some extremely sporting sections.

These are but a few of the more interesting caves and potholes in Yorkshire. The caving regions cover a vast area and the opportunities for further exploration and discovery are by no means exhausted.

WALES

It was not really until after the last war that caving really became a serious sport in Wales. Naturally, caves were known about in the Principality for many centuries but, as in so many other areas, they were places to keep away from because of some local superstition. Arthurian legends have grown up around some of them and he seems to have slept in as many caves as Queen Elizabeth the First apparently did beds. Possibly Celtic imagination and the national flair for romance played no small part in creating the legends and mysteries. Archaeologists and geologists were the first to venture into the Welsh underworld and some of the earliest prints of stalactites and stalagmites were of scenes in the cave on Caldy Island, formerly Ynys Pyr, opposite Tenby in Pembrokeshire. The earliest cave explored in South Wales appears to be at Crawley Rocks, Oxwich Bay, near Swansea, where the remains of elephant, rhinoceros, ox and hyena were found. In *Bosco's Den* at Penard, Glamorgan, hundreds of antlers shed by deer were uncovered. These were truly scientific discoveries. Cave exploration in Wales can be said to begin with the Morgan brothers who discovered *Dan-yr-Ogof* in 1912 and *Glyntawe,* in the Swansea Valley, near to Craig-y-Nos Castle, the home of the celebrated soprano Adelina Patti. The Morgan brothers found the passage above the river which leads into what was later to become a show cave. With considerable courage they tackled two of the four lakes and later opened up part of the cave for public viewing. A later exploration in 1937 by teams of Yorkshire and Somerset cavers pushed on beyond the limits of the Morgan explorations and found another one and a half miles of passages. Later still, in 1965, some cavers who still believed there was more to be discovered in this cave were rewarded for their patience and tenacity. It was to a girl, Eileen Davies, that the privilege fell of breaking through into a new section of the cave. She was successful in forcing an extremely tight and tortuous crawl which was to lead into the magnificent new series. At the time of writing the known length of this cave is now over six miles with the possibility of further extensions.

Nearby is one of Wales' deepest potholes, *Pwll Dwfin,* which is a five pitch climb and reaches a depth of 310 feet with two tricky waterfalls to negotiate on the fourth and fifth pitches.

Two more of the great South Wales systems are still giving up secrets, justifying the many years of work by those who have made these caves their special preserves—Ogof Ffynnon Ddu, also near Glentawe and Agen Allwedd, which is probably better known by its nickname, *Aggie-Aggie*. When I first went into Agen Allwedd, it was little more than 500 yards long. But since its first exploration in 1946 by the South Wales Caving Club, the Hereford Caving Club has made its way into caving annals by thrusting forward to discover mile after mile of new passages. These now extend to more than 14 miles. This is small compared to the sixty miles or so of the Flint Ridge System in America or the fifty-three miles of the Hölloch Cave in Switzerland, but Aggie-Aggie is a great cave system and also needs treating with respect. *Ogof Ffynnon Ddu,* the Cave of the Black Spring, is another long, strenuous cave system, with a stream passage nearly two miles long. To explore it and the new extensions is one of the major cave expeditions in Britain and is not to be undertaken lightly. It is no place for novices and requires the highest standards of ability and training. Provision must be made for the possibility of being cut off for a time as flood conditions seal off some of the sections of the stream passage.

Pant Mawr is another of the best-known Welsh cave systems, which has been extended considerably in the past few years.

All the big South Wales caves require a respectful approach but there is ample scope for all forms of caving as well as for further exploration by digging, diving and patient observation.

North Wales has less spectacular caves to offer although fresh efforts are being made to reawaken interest in this somewhat neglected area. My own club has sparked off some new work but there is nothing, as yet, to compare with the South Wales systems.

The same is also true of the Mid-Wales area and around the Forest of Dean, where there are a number of modest caves which could produce some remarkable results if probed hard enough.

THE MENDIPS

The Mendips claim to possess the oldest stately home in Britain —*Wookey Hole* where rich finds have been made of our early forebears. It was on show to the public long before our landed gentry started opening up their homes to help pay the rent and death duties. As far as exploration is concerned, this is now strictly the preserve of the cave-divers, the only people who can follow the River Axe along the submerged passages well beyond the tourist limits. Here, and at Swildons Hole at Priddy, cave-diving has been brought up to a high standard of perfection.

As in the Peak District and Yorkshire, many of the caves of Mendip were first known to the miners long before archaeological work started. *Lamb Leer* (or Lair) is such a one, being first mined in the seventeenth century. A fine early account of a cave, looked at with a true speleological spirit, is that of Mr John Beaumont, a surgeon who lived nearby at Ston Easton. He published an account of an exploration of Lamb Leer in 1681 in which he described a descent of the seventy-six feet Main Shaft down which he was lowered at the end of a rope. Mining eventually stopped, and for one reason or another Lamb Leer seems to have been 'lost' or forgotten about until 1880 when another company reopened the mine and developed part of it into a show cave. Caving interest revived in 1935, and Lamb Leer remains one of the most fascinating of the Somerset caves.

Swildons is another of the great Mendip systems. It was here that the doyen of Mendip cavers, Herbert Balch, had the frustration of realising its possibilities and even planned an exploration, but missed the first descent in August 1901. He did a great deal there subsequently and was one of the true pioneers of British caving along with Dr Ernest Baker, who, although a Somerset man, started his caving career in Derbyshire at Eldon Hole. From

1901, Swildons has always been a cave in which further discoveries are possible. Also among the Somerset 'classics' are the fairly recent discovery, *G.B. Cave* at Axbridge and *Stoke Lane Slocker*. The latter, although known about for many years, was not developed until 1947 when Pat Browne from King's School, Bruton, who was unfortunately killed on Snowdon in 1951, opened up the way into what is now *Stoke Lane II*.

These are all hard caves and so is the second-largest system in the Mendips—*St Cuthbert's Swallett* at Priddy, which is a veritable maze of passages including one section aptly named the Rabbit Warren. For this reason, the Bristol Exploration Club, whose field HQ is nearby, have adopted a leader system and only allow parties down if accompanied by one of their approved leaders. This may take some of the sport out of the cave for people who like and are capable of finding their own way. In fact this question of access is sometimes a sore point, and it is extensively developed in the Mendips where many of the best caves are under the control (and lock and key) of the major clubs in the area.

Nevertheless, there is ample scope for caving enthusiasts in many parts of the area, notably at Cheddar and Burrington.

DEVON

Devon is usually considered too far away to interest Northern cavers, but it has its own faithful devotees and some extremely active groups. The county has a small number of caves but they are interesting and worth a visit. Some of them were the first to be examined and to establish the co-existence of primitive man with now-extinct mammals. Information about the Devonian caves can be had from the Pengelly Cave Research Centre, named after the eminent geologist William Pengelly FRS, who was born at Torquay. The Centre is at Buckfastleigh and information about its work and the caves of the area may be had from Mrs Averil Longman who is the secretary, at 4 Heath Rise, Brixham.

The Centre is in Higher Kiln Quarry, which is in the middle of

26, 27 Tunnel Cave, Breconshire, South Wales; *(above left)* a large cavern reached by a low crawl; *(above right)* the Main Cavern

28 *(left)* The Escape Route in Ogof Ffynnon Ddu, South Wales

29 A long ramble in a river—that's what Pollnagollum Cave in Eire has been called. This is one of the great cave systems of Ireland, which can absorb a lot of water

30 Flood waters surging along the water passage in Pollnagollum

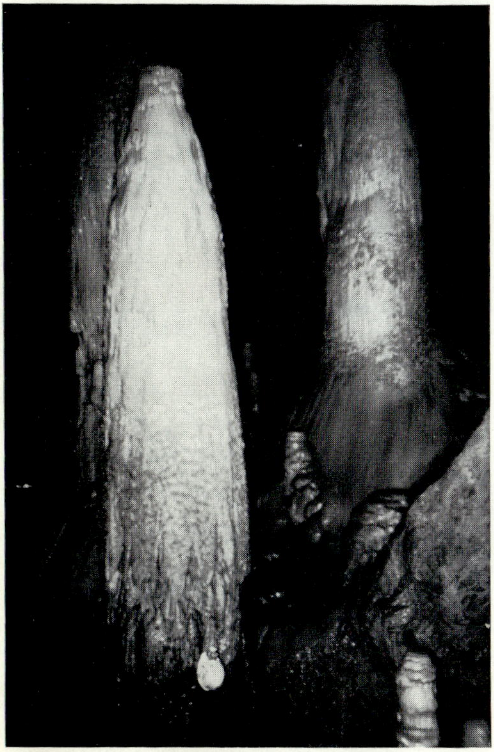

31 (*above*) Large stalactites draping Pant Mawr, Breconshire, South Wales.
32 (*left*) Cave formations range from the delicate and fragile to the immense—like these huge stalagmites in the Gouffre Berger, which dwarf the caver at the foot of them

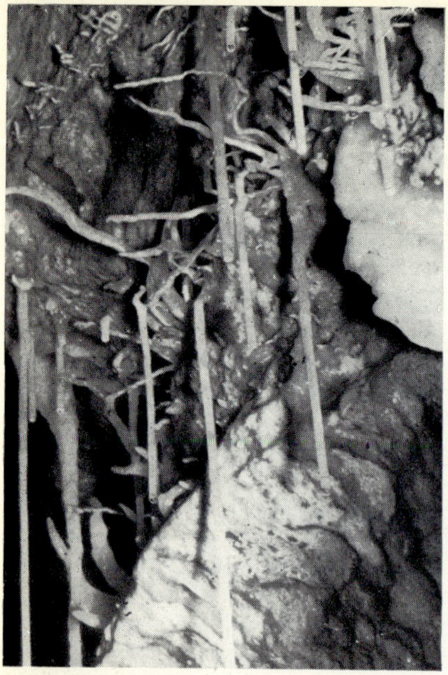

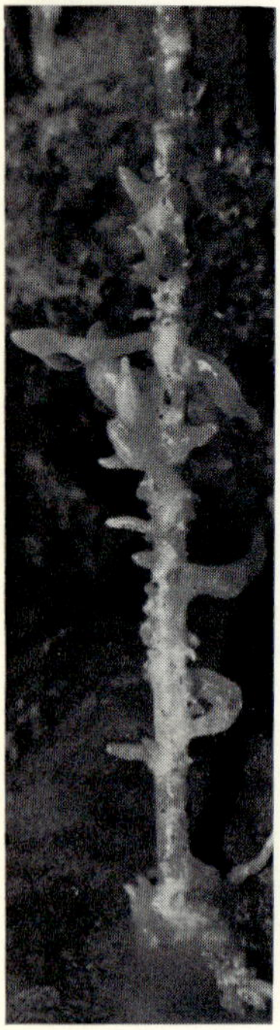

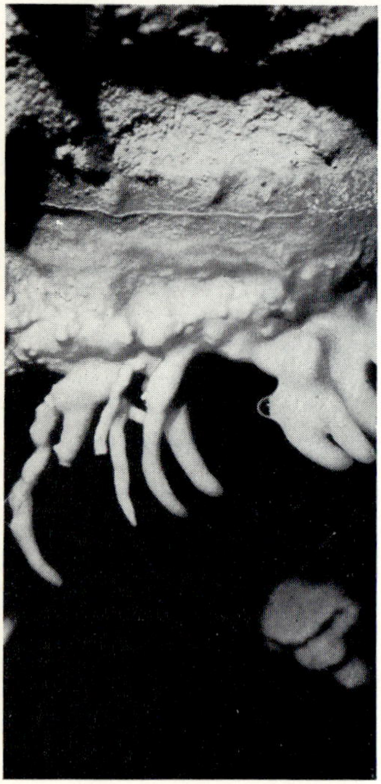

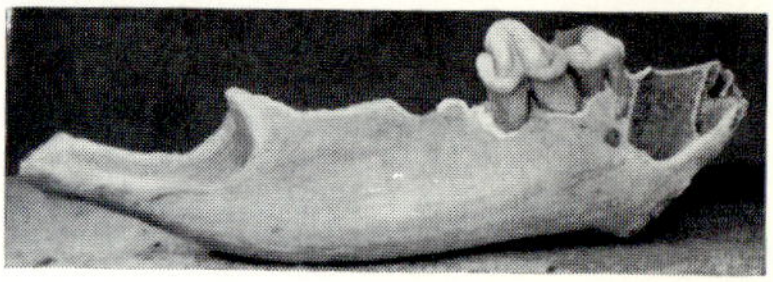

Helictite growths: 33 (*top left*) in Pant Mawr; 35-36 (*left*) in Ogof Ffynnon Ddu. 34 (*top right*) An important find in Derbyshire—human bones from a 'crouch burial' *in situ.* 37 (*above*) The jawbone of a mammoth hyena found in the Elderbush Cave, Manifold Valley, Derbyshire a fairly extensive cave area. There is nothing really 'desperate' by present day standards. Like most of the other caving areas, Devon has its own show caves, with the famous *Kent's Cavern* as the best known. Before the tourists arrived in force, Kent's produced many prehistoric finds.

I think *Pridhamsleigh Cavern* is one of the longest of the Devonshire caves and was once described by a member of my own club as 'the most complicated, arduous and filthiest' he had ever been in. Probably because, at that time, cattle had used the entrance as a shippon and further in there is some of the most glutinous mud to be found in a cave anywhere. Pridhamsleigh offers all kinds of obstacles including a divable 'lake'.

There is a close link between *Bakers Pit Cavern* and *Reed's Cavern,* but I believe no one has yet broken through to complete the round trip. Mr Edgar Reed, who discovered the cave which has his name, told me: 'If you have a long arm, you can shake hands between the two caves.' Reed's Cavern has been called the photographer's ideal cave, because it is easy of access and has some splendid, unsullied formations. Bakers Pit has a complicated and puzzling topography and has repaid many years of work by a breakthrough into another section with some fine formations.

Away from Buckfastleigh, Chudleigh is another caving centre with a number of easy caves. Ladder work in Devon caves is minimal.

FURNESS

I have a great affection for the Furness area of Lancashire, an often neglected peninsular. It has some attractive unspoiled scenery and an interesting geological development which includes a considerable area of limestone. There is not much 'true' caving here and most of the exploration originates from mines. Thanks to the indefatigable work of the Furness Speleological Group, which was formed in 1958, there is now a comprehensive guide to the mines and natural caverns. It is probably the only cave guide which refers to (but does not classify) a railway tunnel at Lindal, where a natural cave was discovered during the building of the tunnel. None of the caves are long except perhaps *Stainton Cave*, which was a modest 800 feet until quarrying lopped part of it away.

While still in the North there are other parts of Lancashire and Cumberland where some prospecting around could bring a number of rewards. The copper mines at Coniston have interested a few cavers and I have done a fair amount of pottering around the Alston area where lead-mining was done on a considerable scale and produced one of Britain's most remarkable waterways, the Nent Force Level, an underground canal four and a half miles long at Nenthead. Around the Cross Fell area there are numbers of mine shafts and from historical reading it seems that some of these broke into natural caverns as well.

SCOTLAND

As yet, Scotland has not figured largely in British caving. The main reason of course, is that the limestone distribution is sparse. Most of the caves in Scotland are sea caves like the *Spar Cave* on the Isle of Skye which has some fairly extensive formations. And way out in Orkney and Shetland there are some good examples, *Braebister* at Hoy and *Papa Stour* off the Shetland mainland, where there is one big chamber at *Kirstan's Hole*. There is also the famous *Ossian's Cave* in Aonach Dubh above Loch Achtriochtan in Glencoe where a block of igneous dyke rock has fallen out.

The limestone caves, such as there are, are confined to Sutherland in the valleys of the Tralligill and Allt Nan Uamh and south-east of Knocham. Recently, discoveries were made in this area which could lead to bigger and better things. Digging and blasting could lead to success. The *Smoo Cave* at Durness has an impressive 100 feet high entrance, a waterfall and a deep lake which requires a rubber dinghy. But there is little else except one main chamber and a few tight passages shooting off.

IRELAND

Cavers who want to go into an area which is still fairly unexploited, where there are possibilities of new finds and important breakthroughs should heed the advice of Horace Greeley, the American journalist: 'Go West, young man, go West.' Ireland, both Ulster and the Republic, has a considerable number of cave systems and I am sure that there are as many more waiting to be found. For a number of reasons, caving is not done on anything like the same scale as in England and Wales. Yet the upland limestone areas of Ireland have a haunting beauty of their own and the underworld is dramatic and colourful. Archaeologists and geologists first probed the dark depths and it is only in comparatively modern times that cavers have been active across the water. Even here can be found the names of the indefatigable Edouard Martel and Ernest Baker. Martel's book, *Irelande et les Cavernes Anglaises,* first published in 1897, included some of the better-known systems and also dramatic sketches of such caves as *Marble Arch* in Fermanagh.

The Irish underworld contains mile after mile of passages and formations to match and, at times, surpass those in other parts of the world. Inevitably, some of the caves figure in legends, songs and poems of old Ireland, and one of the earliest references to a cave is the *Dunmore Cave* in Kilkenny. It was mentioned in AD 928 in the *Annals of the Four Masters,* where Dunmore is described as Derc Ferna which was its ancient name meaning the Cave of the Alders. It is reputedly the scene of a huge slaughter of 1,000 people.

The terminology of Irish caves has the same fascination as those in Wales. The prefix *Poll* abounds like that of *Pwll* in Welsh, both meaning the same thing in general terms—a hole or a cave. It is a little confusing to find there are at least nine or ten Pollnagollums. Pollnagollum means 'cave of the doves'.

There are over 300 known and explored caves in Ireland, dispersed throughout the land in the main caving regions which are in Kilkenny, Laoighis, Tipperary, Kerry, Cork, Waterford, Galway and Mayo in the south, and Sligo, Leitrim, Fermanagh and Cavan in the north and west. These are the areas where limestone mostly abounds and, for the first time, the caves were all conveniently listed by that tireless researcher J. C. Coleman in his book *The Caves of Ireland*. There is only one omission, and that is that there is no classification of the standards of difficulty of the individual caves, although this is no serious handicap and, in some ways, adds to the charm of caving in Ireland. Water passages are frequently found and there is ample scope for lengthy river journeys making a rubber dinghy an essential part of any expedition's equipment.

Caving in Ireland by parties from the United Kingdom must be done on fairly extensive expedition lines as the time factor is one of the problems. A week-end's caving in Ireland is just possible. Once by flying out from Manchester to Dublin, by air from there to Shannon, then travelling by bus to Lisdoonvarna and from there by taxi to the cave, I was able to do a great deal of the *Pollnagollum-Pollelva system*. This is one of Ireland's best-known caves in the Slieve Elva in County Clare. For most cavers, this type of trip would be too expensive. The new ferries make it possible to get over more easily with the vehicles necessary to carry all the tackle required.

It is also worth bearing in mind that cave rescue in Ireland is in its infancy and is nowhere near as well organised as in England and Wales. But the average-sized club expedition should be able to cope with any emergency and should go prepared for any eventuality.

The scope for archaeological expeditions is also excellent in Ireland, but before serious excavations can be made it is necessary to obtain a licence from the Commissioners for Public Works as caves come under the general protection of the National Monuments Acts.

11 *In an emergency*

The growing number of accidents underground led to the creation of cave rescue organisations. They first began functioning before the war when it was realised that to retrieve an injured person from a cave involved the development of special equipment and methods. Today's cave rescue teams are efficient and well-organised, pooling their knowledge not only throughout Britain but internationally as well.

Each caving area has its own rescue teams which, in the event of some major catastrophe, are available to assist each other. These teams, it must be stressed, are entirely voluntary and give their services willingly to retrieve their fellows involved in accidents.

The teams usually operate on a snowball system with a standby team on duty each week-end and the other teams called out if required. The call-outs are in the hands of police divisional headquarters in each area—for example all calls for help in the Derbyshire area, no matter where, are routed to the police at Matlock who speak to the appropriate on-duty rescue controller.

The rescue teams have on hand basic rescue material such as ladders, pulleys, excavation equipment, boulder jacks, special stretchers, pumping and diving gear. They stage regular rescue exercises to gain experience of all the equipment and make a careful study of all the known hazards in the various systems in their area. Many of them are well known, even notorious. Accidents may be due to a genuine slip or mistake, more often than not due to inexperience and foolhardiness. But cave rescuers do not sit in judgement on the pros and cons of whether a person should have done what they did. They get on with the job, no matter how difficult.

In one cave rescue test, I played the part of the 'victim' and came quietly to the conclusion that if anything happened to me underground I would ask for a bottle of rum and a gun. I'd drink the rum, and then shoot myself. Maybe this was because my 'friends' were making sure that I had a not-too-pleasant ride on the stretcher, being dunked under water and pendulumed alarmingly on the ladder pitches. But seriously, cave rescuers are dedicated people. The teams usually include a doctor who is also a caver and is prepared to give up his time both to work on a rescue and also lecture on the care of injured people, when he is not there.

A cave rescue will go something like this—and I am basing this on the experiences of my own caving association and the Derbyshire Cave Rescue Organisation. There may be an accident in the New Oxlow part of the Oxlow Caverns which is a not-too-difficult pothole and invites cavers of moderate ability. A member of the party will rush out and dial 999. Normally, this call would be routed to Buxton and it would be transferred to the County HQ of Derbyshire police at Matlock. If the caller had asked for Cave Rescue, he would immediately have been put through to Matlock, as this is now classed as a regular emergency service.

One of the difficulties of cave rescue starts at this stage—the passing of the correct message. Some cave and pothole names sound alike and Oxlow sounds very much like Oxclose, which is another mine shaft like Oxlow, but a long way off at Snitterton. It could equally well sound like Knotlow Mine, which is in a different direction altogether. The person passing an emergency message must therefore speak absolutely clearly. In some circumstances, such a mix-up might be funny—but not in a cave rescue.

At Matlock there is a cave rescue information book which indicates which DCRO controller should be contacted. Usually it is nearest to the accident—but he may be away. There are always others on duty. With the accident being at Oxlow, the controller would be at Buxton or Sheffield. The teams living nearest to the accident or the standby team will be called out and others put on standby. At the scene of the rescue, there will be two controllers—one underground to supervise the rescue work there and one above ground, working with the police and co-ordinating the above-ground services.

The first to arrive are sent underground to evaluate the situation and the stretcher and other equipment follows. A doctor goes in and so does a telephone-laying team of two potholers to provide the vital link between the scene of the rescue and the above-ground services. If the rescue looks like being a long one, the teams further away are contacted.

The victim is put into an exposure bag designed to prevent loss of body heat and the newly converted Neil Robertson stretcher will be used in conjunction with a drag stretcher of thick rubber on which the injured person can be pulled along. Meanwhile, the cave may be re-rigged with the DCRO's own ladders, pulleys and hauling ropes, which may be stronger and more reliable than those of the original party. Oxlow is not as bad as the tight, extremely difficult Giants Hole or Nettle Pot.

The stretcher would be counter-balanced up in Oxlow. The underground controller would be in constant contact with the surface and extra equipment or medical supplies would be sent down as required. If all goes well, a rescue in New Oxlow could take six to eight hours and involve, in all, thirty men, but others—as in Giants Hole or when a young soldier boy disappeared in Carlswark and was found alive by a remarkable piece of good fortune in the only air space there was after diving—may take a hundred or so cavers.

For those rescues which require specialised equipment for pumping or drilling there are manufacturers who are only too willing to lend it and this is not peculiar to Derbyshire.

I have gone quickly over this ground of a rescue to remind all cavers of what is involved. My friend John Needham has pointed out: 'Though the rescue organisation is efficient, don't forget the next time you decide to climb a ladder without a lifeline it takes a rescue team a long time to reach you. If you are two hours in a cave, it will take a man two hours to take out a message, at least one hour for the team to turn out, and two hours more to reach you—at least five hours. It just isn't worth letting the standard of your caving relax. The hardest task in any rescue is to be the victim.'

But people do relax their standards or do not have high enough standards, so it is useful to learn the basic procedure should you or one of your party be involved. This has been well laid down by the Mountain Rescue Committee with the help of many caving experts.

If a fall occurs try to reach the casualty at once as, if he is lying unconscious near water or in water, he could drown. As I have pointed out elsewhere, shock and exposure are more likely to be fatal than further physical damage caused by moving him. First aid should be given (there should always be someone trained or with knowledge in first aid in every caving party). An assessment of the injury should be made. If in water, the casualty must be moved out of it and also away from strong draughts which may chill him. Keep him warm.

Physical exhaustion, cold, and wet can collectively be fatal. A man suffering from them slows down and wishes to sleep. Food, rest and warmth are needed, but the casualty must be allowed to use what strength he may have to get somewhere comfortable.

Borrow extra clothes from other members of the party; warmth can also be communicated by other members of the party huddling close. Try to insulate him from the cold of the cave floor and persuade him to eat, if possible. Candles, if they can be spared, should be used also to give a little extra heat as well as psychological comfort. One or more competent persons who are fully conversant with the injuries and the situation of the casualty should go out to summon aid—preferably two people as a man going out on his own could also become a casualty.

No matter what happens, a casualty should never be left alone. If the cave is not too difficult and the casualty can help himself somewhat, start to move him towards the entrance before the rescue party arrives as this will cut the time taken to get him out.

Do not panic. This can lead to mistakes and mishearings of a cave location as I have pointed out. A person ringing for help should give the location of the phone box, or the telephone subscriber's address if it is from a farm or private house.

The whole success of a rescue depends in many ways on the people originally with the casualty doing the right things and not aggravating the situation. An emergency calls for clear thinking, some aspects of which I refer to in Chapter 8.

Appendix I: A CAVING SAFETY CODE

1 Never go caving without leaving full information about which cave or pothole you are exploring. If in camp or in a vehicle leave a card with your expedition clearly marked and your estimated time of arrival back at base. Never change your mind and go somewhere else.

2 Always go caving with adequate clothes, boots, helmets and good lights. Try and keep the upper part of the body dry with waterproof clothing.

3 Always be adequately fed and take food including glucose for quick energy. Never take alcohol into a pothole and never under any circumstances give it to an injured person.

4 Use safe, approved tackle. Always use a lifeline on ladders no matter how short the pitch may be. Guard against ropes being trodden on or affected by acid. Check your equipment regularly.

5 If a person is in difficulties on a pitch or has a lifeline jammed or twisted behind the ladder and he does not have a bandoleer or chest harness, lower a looped rope to work over his shoulders and under the arms.

6 If someone on a ladder falls he may be incapacitated temporarily due to being winded or losing consciousness. If you know there is a safe point below, lower him—unless the man on the safety line tells you not to do so.

7 When protecting a potholer on a ladder with a lifeline, make sure you have a firm stance and always be belayed to a secure point, if possible. If there is no rock to which you can fix an anchorage, insert a piton. Watch the running out of the rope and listen carefully for any whistle signals or shouts from the man on

the ladder. Try and insure that the rope does not go over sharp edges of rock and use the correct belay technique to absorb the shock of a fall.

8 Take great care not to disturb loose stones and boulders while standing on ledges or directly above other members of a party. Do not shout in unstable rock conditions. Do not forget that only one member of a party should move at a time. Others in the party should not move until it is safe to do so.

9 If your light fails, stay put and do not move if you cannot see. Use your emergency lighting or wait until other members of the party join you.

10 Never attempt to swim underground fully clothed, except when protected by a safety line. The weight of wet clothing and boots will pull you down.

11 Check carefully if the cave system you are entering is liable to submergence due to flooding. Before entering such a cave check all weather reports and study local conditions at the time of the descent. Remember that in winter a thaw can cause a sudden rise in water levels.

12 In confined spaces be wary of an oxygen lack if wearing an acetylene lamp which may burn it up. Switch to electric lighting if you have it. Move slowly and economically so as not to lose strength. And never go into such a confined space unless someone is near at hand to help should you be overcome.

Appendix II: CAVE RESCUE

In the event of an accident, the first contact should be with the police either by dialling 999 or the police number if it is known. They will call out the stand-by rescue team and the surface controller will call out the others, if needed.

YORKSHIRE

Police: Ingleton Clapham area, Settle 2142; Grassington area, Grassington 222;

Upper Wharfedale Fell Rescue Association: Mr L. H. Huff, Station House, Grassington, Skipton; *Cave Rescue Organisation:* Mr Reg Hainsworth, Sunnydale, Ingleton, via Carnforth.

PEAK DISTRICT

Police: Matlock 3131 or Buxton 2811;

Derbyshire Cave Rescue Organisation: Mr John K. Needham, 67 Nunsfield Road, Buxton.

NORTH-EAST ENGLAND

Police: Stanhope Section Office, Stanhope, 205; Consett Divisional HQ, Consett 4204;

Durham Team Cave Rescue Organisation: Mr J. Newrick, High Whitestones, Ireshopeburn, Bishop Auckland.

SOUTH WALES

Police: Glamorgan County Police HQ, Neath 3964; Bridgend 2404;

South Wales Cave Rescue Organisation: Mr G. L. Clissold;

South Wales Caving Club: 'Silhouette', Staunton Coleford, Gloucester.

WEST OF ENGLAND

Mendip Area

Police: Bristol 22022; Somerset and Bath, Taunton 7911; Frome Divisional HQ, Wells 3481;

Mendip Cave Rescue Organisation: Dr Oliver Lloyd, Withey House, Withey Close West, Bristol 9.

Devon Area

Police: Newton Abbot Division HQ, Ashburton 210;

Devon Cave Rescue Organisation: Mr E. Reed, 12 Market Street, Buckfastleigh.

IRELAND

Police: Belfast 650301;

Irish Cave Rescue Organisation: Miss F. Tinsdale, 21A Thomas Street, Portadown, County Armagh.

Appendix III: CAVE PRESERVATION CODE

1 Cavers, like all other people who go out of doors, should have the goodwill of the farmers and country communities at heart. Care should be taken to obtain permission to cross private land and cavers should co-operate with farmers, landowners and their agents, because thoughtless actions by cavers prevent other parties from enjoying the pleasures of underground exploration. Cavers should uphold the Countryside Code and guard against risk of fires, fasten all gates through which they pass, keep dogs (if they have them) under control and keep to the paths across farmland. It is important not to trample on growing crops. It should be remembered, too, that grass is also a valuable growing crop and cannot be cut by machinery if flattened. Avoid damaging fences, hedges and walls.

2 Make no excavations or diversions of water without permission. If fluorescein is to be used check if it is going to go into the drinking water of a farm or a village community. Warn people in advance of its use and stress that it is quite harmless. If excavations or diversions of water are made seek permission and make sure that adequate care is taken to prevent animals and people from falling in.

3 Use adequate lighting so that fragile formations and other interesting features of the underworld, particularly the crystalline floors and roof formations, are not disfigured with mud or destroyed by colliding with them.

4 Choose a route carefully and make no additional tracks unless it is absolutely necessary.

5 Leave stalactites, stalagmites, helictites, flowstone and rimstone pools as clean and undamaged as when they were found.

6 Commit no acts of vandalism by breaking off portions of formations to take out as souvenirs. Remember other people want to see them *in situ*.

7 Never disfigure the walls of a cave by writing or drawing on

them. Stop people from blackening their initials on cave walls with carbide lights or candles. If a record must be left of your depth in a cave, use metal club tags and leave them.

8 Try to leave the natural life of a cave as undisturbed as possible. If specimens are required take only enough for identification.

9 Do not disturb the evidence of floods or other episodes in the cave's history.

10 Treat the cave as a Nature Reserve. Avoid polluting the water by various means. Take out your spent carbide and batteries.

11 If there are signs of human or animal occupation do not disturb them until they can be examined by experts.

12 Let someone small tackle the exploration of a passage where formations might be damaged.

13 If it is necessary to drain pools, use wherever possible the siphoning method, so as not to upset the natural conditions of a cave. Avoid, if possible, permanently withdrawing water from a part of the system as this can upset the natural balance of a cave.

14 Keep caves tidy and leave no litter. If you find any, take it out with you.

15 Be careful when taking photographs, especially when using flash powder which has explosive qualities and can damage delicate formations.

16 Leave the cave and its contents in the condition you would expect to find it.

17 Respect other cavers' digs and explorations and pay such dues as may be required.

18 Discourage solo descents and ensure that members of your group are properly trained.

19 Be courteous and helpful when dealing with newspaper, television and broadcasting reporters. Remember they have a job to do and require only the facts for their reports.

Glossary of caving terms

Abseil: A method of roping down a pitch instead of using a ladder. Also known as a rappel.

Active system: A cave system which is still in the course of formation.

Adit: A word used in lead-mining to indicate a level or sloping entrance to a mine. Often a drainage tunnel from a mine.

Affluent: A tributary system of a river, stream or lake.

Afflux: The upstream rise of water level above normal surface of water in a channel or passage due to contraction or obstruction of the normal waterway.

Afterglow: The fluorescent reaction of some formations which glow, usually green, after a strong light such as a flash bulb or torch is shone on them. An indication of a younger formation.

Albinism: Animals, fish, plants and insects lacking normal colouration due to deprivation of sunlight.

Alluvium: Finely grained silty material deposited by a river or stream.

Ameliorite: A form of crystalline stalagmite more popularly called cave flowers.

Anastomosis: A branching network, like arteries, of tubular channels in limestone, usually in bedding planes; rarely in joint planes.

Anemolite: A stalactite (often called a helictite) which forms contrary to gravity, which as its name indicates is formed by the passage of air or wind curving it away from the vertical.

Anthodite: Another form of cave flower formed from long, fine crystals of gypsum or aragonite.

Aquifer: A belt of waterlogged permeable rocks.

Aragonite: Calcium carbonate in crystalline form. It is slightly heavier than calcite.

Aven: A shaft in the roof of a cave leading to an upper passage.

Beck: An old Norse word often used in Yorkshire and the Lake District for ' stream '.

Beck head: The emergence of a stream at its exit point from underground.

Bed: A layer, deposit or stratum of rock, usually in a belt of sedimentary rock.

Bedding cave: A cave formed by the erosion of a bedding plane —usually low and wide.

Bedding plane (or stratification plane): The separating point between one stratum of rock and another.

Bedrock: The ultimate rock on which alluvial or detrital material rests.

Belay: A point to which ladders or rope anchorages can be fixed. Also the use of ropes for the protection of cavers and climbers.

Blende: Zinc sulphide, sometimes known as sphalerite, zinc blende or Black Jack. Occurs with galena in various deposits, and replaces calcite in limestones.

Blowhole: Sometimes known as a gloop—a small hole in the floor or ceiling of a cave through which a draught is detected. More usual in sea caves.

Blue John: Fluorspar, fluorite or Derbyshire spar. Calcium fluoride found mainly in Derbyshire where it is worked ornamentally, formally as the flux in the steel trade.

Bone cave: A cave in which bones, usually prehistoric, have been found.

Bore: A fast-flowing river with great erosive power due to its speed and the chiselling effect of stones carried in it.

Bore hole: A hole bored by the pressure of water. A hole bored in the floor of a cave to take samples or detect a further cave or passage beneath.

Boss: A short, stubby, rounded stalagmite.

Boulder choke: A blockage caused by rock fall from roof to ceiling.

Boulder clay (or till): A clayey deposit containing stones and erratic boulders.

Bowline: A knot used for belaying.

Brachiopod: A fossilised shellfish known as a lampshell found in carboniferous limestone.

Breccia: A group of rocks consisting of angular fragments. They may be formed by crushing or faulting, volcanism or erosion. Cave breccia often includes fossil bones.

Bridge: A rock arch, or a rock wedged across a passage. Also a technique of climbing by spanning a crack or chimney.

Cache: A storage place for provisions for exploration and emergency.

Calcareous: Containing, or formed from, calcite.

Calcite: Calcium carbonate $CaCO_3$. White to yellowish in colour—the main constituent of limestone and cave formations.

Canal: A cave passage of still or static water.

Capillary action: The movement of water in the interstices of rock due to capillary forces.

Capillary fringe: The zone immediately above the water table containing capillary water.

Carboniferous: A geological period in which the lower carboniferous limestones were laid down.

Cascade: Calcite formation which has coated rocks and looks like a ' frozen ' waterfall.

Cat hole: A very tight squeeze through a very small crack or hole in the rock.

Cat-run (or walk): A very tight crawl.

Cave: Any underground hollow or natural cavity travelling in a horizontal direction. A cave may lead to a pothole—a pothole may include a cave system. Usually a system which does not require the use of ladders and ropes to explore.

Cave art: Paintings, engravings, statues made by Palaeolithic Man.

Cave deposit: The debris on the floor of a cave consisting of clay, silt, gravel, residue of formations.

Cave of debouchure: A cave entered at the opposite end from which a stream enters.

Cave pearl: A round piece of calcite formation resembling a real pearl created by being rotated by water action in a pool.

Cave-system: The whole complex of interlinking passages underground.

Caver: A cave explorer or speleologist.

Cavern: Generally taken to mean a large underground cave or system.

Cavernicolous: A form of life which fulfils its complete cycle underground.

Cavernophilous: Forms of life which exist temporarily underground.

Cerussite: White lead ore and often associated with galena.

Chamber: A large cavern.

Chert: Beds and nodular concretions in limestone.

Chimney: A narrow, vertical fissure climbed by jamming.

Chockstone: A rock wedged in a narrow cleft, sometimes used as a belay point.

Choke: A blockage of a passage, chimney or pothole by rocks, clay, silt, etc.

Claustrophobia: A morbid fear of enclosed spaces.

Clinometer: An instrument for measuring the dip of a stratum or fault.

Clints: Fretting of surface limestone blocks caused by corrosive action.

Column: A stalactite and stalagmite which have bonded to form a pillar.

Conglomerate: A consolidated rock composed of gravel or pebbly deposits.

Controlled entry: A system of allowing access to caves only with permission by barring the way in with lids, doors and locks.

Corrasion: The abrasion of a stream passage by rock held in suspension in water.

Corrosion: The chemical work performed by water containing carbon dioxide in eroding rock.

Crack: A small fissure.

Crawl: A low passage along which it is only possible to proceed on hands and knees.

Crinoidea: The most familiar fossils in limestone; sometimes stone lilies.

Cross fault: A subsidiary fault, almost at right angles to a main fault.

Crystal pool: A pool with little or no overflow which contains crystal deposits.

Current marks: Referred to variously as scalloping or fluting. Marks caused by water power rushing against the rock.

Curtain: Sometimes refers to a row of stalactites hanging down in a horizontal line. More usually it means a thin sheet of formation (streaky bacon) hanging down from the roof of a cave.

Dark zone: That part of a cave which is beyond the limit of daylight.

Deads: Barren veins of rock. Often used by miners of old to build protective walls.

Debouchure: The resurgence point of an underground stream where it emerges from the cave mouth.

Detritus: Rock, sand or gravel deposit formed by river action.

Dig: A place where a cave explorer digs to remove a blockage to further exploration or to find a new cave.

Dip: The angle of inclination of a bed of rock from the horizontal, indicated in degrees.

Distributary: A branch of a stream flowing from the main stream but not rejoining it again.

Dog tooth crystal: A variety of calcite with crystals showing a combination of scalenohedral and prism.

Doline: Surface depressions caused by the collapse of large caves below.

Dolomite: Calcium and magnesium carbonate ($CaCO_3$, $MgCO_3$). Occurs in widespread sedimentary beds. Named after the French geologist, Dolomieu.

Drag stretcher: A specially designed stretcher made for cave rescue. It has eyelets and attachments for dragging in low crawls and is usually made from rubber.

Drainage level: A level tunnel driven by miners to drain a mine.

Drapery: Curtain formation which is in waves or folds.

Dripstone: A general word for formations caused by falling water with chemical in suspension.

Duck: A point where the roof of a cave meets the water level and can be traversed by a quick swim on lung power or by ducking under and coming up at the other side. Often misnamed a siphon. Cave-divers also refer to a duck where there is a minimal air space and they must traverse completely submerged.

Ebbing and flowing pools: Pools where water rises and falls rhythmically, possibly by syphonic action.

Effluent cave: Cave of debouchure.

Electron ladder: A ladder made of wire with metal rungs, light to carry, easy to instal and eliminating the use of moisture-holding rope ladders.

Encrustation: A crust of formation covering anything in a cave.

Entomology: The study of insects.

Erosion: The wearing away of rocks by chemical or other physical means.

Exposure: The deterioration of a body due to wet, cold and inadequate amount of food and liquid. Also used to refer to the steepness of a rock pitch and the depth one could fall.

Exposure suit: Specially designed clothing to keep a body warm and dry and to combat cold in adverse cave conditions.

Eye splice: A loop made on the ends of electron ladders to facilitate clipping to further sections and belay points.

Fault: A fracture in the rocks or earth's crust displacing one side relative to the other.

Firing: Exploding a charge to loosen rock.

Flattener: A low crawl in a bedding plane.

Flowstone: A continuous covering of calcite on a floor or wall.

Fluorescein: A powerful, harmless dye used for tracing the underground flow of water and its emergence points.

Fluorspar: See Blue John.

Formation: Generally, all forms of calcareous deposits in caves creating the various shapes, and also takes in aragonite and gypsum.

Fossils: The preserved remains of life from previous geological ages.

Galena: Lead sulphide. An important source of lead. Extensively mined in limestone and sandstone.

Gallery: A high-level cave passage linking one part of the system with another part of the main system.

Ginging: Lining a lead mine or other shaft to prevent the sides caving in.

Gloop: See Blowhole.

Gour: A rimstone pool.

Grike: The fissure created in clints by the solution of rock in the joints.

Ground water: Water below the water table.

Grotto: A picturesque cave, usually lit up in show caverns.

Gypsum: Hydrated calcium sulphate ($CaSO_4 . 2H_2O$).

Hade: The angle made by a fault or fissure in relation to the vertical. *See also* Dip.

Header: Top extension of a ladder used for joining to another section, or securing at the head of a pitch.

Helictite: See Anemolite.

Heligmite: A stalagmite growing away from the vertical and often changing the direction of its growth.

Hydrology: The scientific study of water and water passages underground.

Hypogean: The zones below the endogean or surface zones.

Ice Age: The last period when Britain was covered with ice.

Interstices: Small spaces or cracks in the rock.

Joint: The natural division of rock, other than a bedding plane, usually at right angles to it. A joint may be confined to one bed or traverse several beds when it is known as a master joint.

Karabiner: A metal snap ring, in various shapes with various locking devices, for fastening to ropes, linking ladders, using on belays.

Karst: A Yugoslav area which has given its name to the general characteristics of limestone country.

Knee wrecker: Any crawl with accretions on the floor painful to traverse unless the knees are protected with pads.

Labyrinth: A maze or complex of cave passages.

Ledgements: Wooden platforms in mine workings.

Lench: A landing place in a mine.

Letterbox: A variation on a cat hole; a narrow slit in the rock.

Level: A horizontal gallery in a mine.

Lifeline: The safety rope attached to a caver negotiating a pitch to safeguard against a fall or to assist in the ascent.

Limestone: The general term given for the bedded rock composed essentially of calcium carbonate, soluble in weak solutions of carbon dioxide and sulphur dioxide in rain water.

Master cave: The main passage in a system into which all other passages eventually converge.

Maypole: A scaling pole, often jointed, to reach higher passages.

Meander: A winding cleft or passage in a cave.

Moon or mountain milk: Calcite in a creamy, liquid form or like cheese which has not set.

Ogof: Welsh word for cave.

Oolith: Ellipsoidal or spherical bodies composed of concentric layers of calcite, aragonite or iron carbonate. Small cave pearls.

Ooze: A deposit of calcareous and siliceous skeletons of sea organisms.

Oxbow: A loop passage which formerly contained water.

Permeability: The ability of water to pass through pores in the rock, joints and bedding planes.

pH value: A convenient method of expressing differences in acidity or alkalinity of a solution in factors between 0 and 14, and is a measure of the hydrogen-ion content. A pH value of 7 is the neutral point. Values above 7 are alkaline; below 7 are acid.

Phreatic zone: Waterlogged zone in an area of permeable rock.

Pisolith: See Cave pearl.

Pitch: A vertical section of a cave which requires laddering. Cave guide books give the measurements in feet of the different pitches, the amount of ladder and lifeline required.

Piton: A metal peg used as a means of securing a ladder; providing a belaying point.

Pothole: Generally, any vertical shaft. It can be a direct entrance into the limestone, or it may be found in a cave.

Reptation: A word used by French cavers to indicate travelling like a reptile in a confined space, sometimes proceeding naked and dragging one's equipment fastened to one's ankle, so as to be able to manoeuvre in the tightest possible space.

Resurgence: The emergence of water from an underground stream at the surface.

Rimstone: Formation which has grown to form the edge of a pool, deposited by the overflow water (*See* Gour).

Rising: A point where water from an underground source rises to the surface.

Rock milk: A soft form of calcite (*See* Moon or mountain milk).

Rock mill: A hole in a stream bed made by stones spun around by the water force.

Ruckle: A mass of jumbled boulders.

Rudaceous deposits: Sometimes known as psephites. Unconsolidated rock fragments of gravel, scree or boulder clay.

Run: A branch of a vein in a Derbyshire lead mine.

Saturated: Waterlogged.

Scalenohedral crystal: See Dog tooth crystal.

Scallop: See Current marks.

Selenite: Crystallised variety of gypsum.

Shaft: The vertical entrance to a mine; also a pitch of a pothole.

Shakehole: A depression of the surface in limestone country indicating the collapse of a cave or a pothole beneath.

Sink or sinkhole: A point where water drops underground or has previously done so.

Siphon: A passage where water flows by simple siphonic action.

Slickensides: The polish and grooves caused by friction and movement of one rock face against another.

Slocker: A Mendip term for pothole or swallet.

Speleology (or spelaeology): The science of cave exploration.

Spelunker: The American name for a speleologist.

Spongework: Raw materials for chert layers in some limestones. A mass of tubular passages, which look like a sponge.

Squeeze: A tight passage through which it is just possible to wriggle one's body.

Stalactite: Formations suspended from the roof of a cave, mainly formed from calcite.

Stalacto-stalagmite: American term for what is often called in Britain a column. Where a stalactite and stalagmite bond together.

Stalagmite: The reverse of a stalactite, i.e. columns of calcite extending upwards from a cave floor.

Stemple: A lead-mining term for wood or metal wedged across a narrow passage in the form of a ladder, or a place for hoisting something.

Straw stalactite: A stalactite, thin-walled and regular in shape like a straw.

Strike: The horizontal line of a bedding plane at right angles to the true dip; the direction along which an ore body runs.

Sump: The point in a cave where water and rock meet, beyond which point no exploration has been done. Alternatively a water trap.

Swallet: Openings in limestone into which streams disappear underground.

Tail: End of a rope or electron ladder beyond the last rung—opposite of header.

Threshold: Entrance to a cave or pothole to the ultimate limit of natural light.

Trap: See Sump.

Troglobite: The biological name for creatures living permanently in the dark or hypogean areas of caves.

Troglodyte: A human cave-dweller.

Troglophile: A creature which may live permanently in the dark zone of a cave, but which may also inhabit the endogean areas as well.

Trogloxene: A creature which may enter a cave, e.g. a bat, but does not live there permanently.

Tube: An almost circular cave passage.

Unconformity: A break in the sequence of rocks in which beds repose on other beds out of precedence in their geological succession.

Vadose zone: The zone above the water table through which water seeps downwards to the zone of saturation.

Water table: The upper limit of a phreatic zone or the zone of saturation.

Wet suit: An anti-exposure suit made from neoprene designed to keep out cold and damp; made in thicknesses of one-eighth or three-eighths of an inch.

Caving clubs in Great Britain

This list of caving clubs is as up-to-date as possible and I am grateful to the British Speleological Association and the Derbyshire Caving Association for their assistance in its compilation.

Abertillery Youth Caving Club:
10 Cross Street, Abertillery, Monmouthshire

Aberystwyth University C.C.:
24 Clive Road, Puttinghum, Wolverhampton

Accrington Caving Club:
c/o Abbey Hotel, Black Abbey Street, Accrington, Lancs.

Aldermaston Mountaineering Club:
Caving Section, M. R. Wall, 505 D3., A.W.R.E., Aldermaston, Berks

Arborfield Army Apprentices School:
Adventure Group, Capt. J. D. Churchill, Arborfield

Ashford Speleological Society:
R. O. Gilson, Naphill, High Wycombe, Bucks

Bedford School Caving Club:
Sanderson's, Rothsay Place, Bedford

Beechen Cliff Speleological Society:
8 Wiltshire Way, Fairfield Park, Bath

Beeston Mines Research Group:
Lucknow, 3 Humber Road, Beeston, Notts.

Berkeley Mountaineering & Caving Club:
17 Frome Court, Thornbury, Bristol

Berkeley Nuclear Laboratories Caving Group:
27 Midland Road, Stonehouse, Glos.

Birmingham Enterprise Caving Club:
14 Grafton Road, Birmingham, 11

Birmingham University Speleological Society:
The Union, The University, Edgbaston, Birmingham

Bolton & District Pothole Group:
59 Stonesteads Way, Bromley Cross, Bolton, Lancs.

Bolton Speleo Club:
36 Norwood Grove, Bolton, Lancs.

Border Caving Club:
4 Christine Close, Ash, Aldershot, Hants.

Bowline Caving & Climbing Group:
28 Mount Street, Tipton, Staffs.

Bradford-on-Avon YHA Group:
5 St. Margaret's Place, Bradford-on-Avon

Bradford Institute of Technology Pothole Club:

Gt. Horton Road, Bradford, 7, Yorks.
Bradford Pothole Club:
7 Lysker Drive, Leeds Road, Otley, Yorks.
Bredon School Speleological Society:
Bredon School, Pull Court, Bushley, Nr. Tewkesbury, Glos.
Brighouse & District Rover Crew, Caving Section:
8 Newton Park, Hore Edge, Brighouse, Yorks.
Bristol Boy Scouts Caving Section:
1 St. Edward's Road, Church Lane, Bristol 8
Bristol Cathedral School Caving Club:
c/o Cathedral School, College Green, Bristol 1
Bristol Exploration Club:
699 Wells Road, Bristol 4
Bristol Grammar School Caving Club:
Bristol Grammar School, Bristol 8
Bristol University Speleological Society:
The University, Bristol 8
British Aircraft Corp. Sports & Social Potholing Section:
Warton Aerodrome, Preston, Lancs.
British Speleological Association:
Ian Standing, Grove Cottage, Watledge, Nailsworth, Glos.
British Nylon Spinners Speleology Section:
18 Tymynydd, Pontnewydd, Cwmbran
Brooklands Speleological Society:
62 Annett Road, Walton-on-Thames
Brynmawr Caving Club:
124 King Street, Brynmawr, Brecon
Burnley Caving Club:
57 Reed Street, Burnley, Lancs.

Cadoux Caving Club:
25 Kipling Road, Derker, Oldham, Lancs.
Caldicot Caving Club:
35 Firs Road, Caldicot, Chepstow, Mon.
Cambrian Caving Association:
29 Wrekin Road, Wylde Green, Sutton Coldfield
Cambridge University Caving Club:
F. Morland, Clare College, Cambridge
Camerton Excavation Club:
19 Portland Place, Bath, Somerset
Cardiff C.D. Association Caving Club:
56 Gabalfa Avenue, Gabalfa, Cardiff
Cardiff University Caving Club:
Mrs M. A. Williams, Dept. of Microbiology, UCW *and* MON, Cathays Park, Cardiff
Cave and Crag Club:
154 Station Road, Wylde Green, Sutton Coldfield
Cave Diving Group, The:
7 New Street, Wells, Somerset
Cave Diving Group:
17 Sandown Crescent, Gorton, Manchester 18
Cave Diving Group:
c/o B.S.A., H.Q., Settle, Yorks,
Cave Photographic Group:
62 Bradford Lane, Bradford, Yorks.
Cave Rescue Council:
John Plowes, Malt Kiln Lane, Elsham, Brigg, Lincs.
Cave Research Group:
Arthur Butcher, 369 Stone Road, Trentham, Stoke-on-Trent
Cerberus Spelaeological Society:
H J. Hill, Squares Road, Chilton Trinity, Bridgwater, Somerset
Chelsea Spelaeological Society:
12 Heybridge Drive, Barkingside, Ilford, Essex
Cheltenham E.I. Northlands Youth Centre Caving Section:
7 Firnleigh Crescent, Coomb Glen, Cheltenham

City & County of Bristol OB Scouts Association:
1 St Edward's Road,
Clifton Wood, Bristol 8

Cotham Caving Group:
40 Thornleigh Road, Horfield,
Bristol 7

County Caving Club:
14 Blenheim Road, Six Bells,
Abertillery, Mon.

Craven Pothole Club:
The Annexe, Beech Hill,
Carleton, Skipton, Yorks.

Crewe Cave & Pothole Club:
28 Hall O'Shaw Street,
Crewe, Cheshire

Croydon Caving Club:
S. M. Norman, 'Madras',
Bedwardine Road,
London, E 19

Croydon Scout Speleological Group:
68 Harold Road, Upper Norwood,
London, SW 19

Dakotas Clifton Caving Club:
25 Musters Road, West
Bridgford, Nottingham

Derbyshire Caving Association:
Mrs P. E. Lunn, 28 Kenbourne
Road, Sheffield

Derbyshire Caving Club:
Arthur Ball, 4 Charlotte Street,
Cheadle, Cheshire

Derbyshire Pennine Club:
68 Slayleigh Lane,
Sheffield 10

Derbyshire Speleological Group:
5 Dovecliffe Crescent,
Stretton, Burton-on-Trent

Devon Spelaeological Society:
15 South Avenue, Heavitree,
Exeter

The Distillers Co. Ltd Caving Club:
G. F. Batty, Hed-Fan, Sully
Road, Penarth, Glam.

Dudley Cave Rescue Team:
C. W. Taylor, 1 Myrtle Road,
Priory Estate, Dudley, Worcs.

Dudley Caving Group:
28 Mount Street, Tipton, Staffs.

Durham Cave Club:
Clandown, 8 Darlington Road,
Neville's Cross, Durham City

Dursley Caving Club:
The Rectory, Dursley, Glos.

Earby Pothole Club:
51 Harrison Street, Barnoldwick,
via Colne, Lancs.

East Devon Caving Group:
Stanley Villa, Crewkerne Road,
Chard, Somerset

East Somerset Caving Club:
Vernigore Cottage, Welham,
Castle Cary, Somerset

Ebbw Vale Youth Caving Club:
102 Tredegar Road,
Ebbw Vale, Mon.

Eccles Caving Club:
B. Saville, 8 Park Drive, Monton,
Eccles, Manchester

Eldon Pothole Club:
R. H. Whitehouse, Hall Close,
Cossington, Leics.

Elton Speleological Group:
60 Goodland Street, Woolfold,
Bury, Lancs.

Exeter University Speleological Society:
Heatherington House, Duryard
Halls, Cowley Bridge Road,
Exeter, Devon

First Abertillery Senior Scouts:
24 Caen Street, Abertillery, Mon.

First Nottingham YMCA Rover Scouts Potholing Section:
34 Birell Road, Sherwood Rise,
Nottingham

First Panteg Senior Scouts Caving Group:
Sunnybank, Sluvad Road,
Panteg, Pontypool

First Slough Rover Crew:
29 Iona Crescent, Cippenham,
Slough, Bucks

Fylde Mountaineering Club:
Potholing Section, 11 Dodgeons
Close, Poulton-le-Fylde, Lancs.

Giggleswick School Potholing Club:
Giggleswick School, Nr. Settle,
Yorks.

Glantaff Caving Club:
24 Prospect Drive, Llandaff, Cardiff
Glasgow Speleological Society:
120 Lochlea Road, Newlands, Glasgow, S 3
Gloucester Speleological Society:
J. V. Elliott, 71 Imjim Road, Priors Farm, Cheltenham
Grampian Speleological Group:
Alan Jeffries, 8 Scone Gardens, Edinburgh 8
Gritstone Club:
22 The Rise, Morris Lane, Leeds 5
Gritstone Tigers:
P. Mantle, 122 Logan Street, Market Harborough, Leics.
Gwent Caving Club:
14 Blenheim Road, Six Bells, Abertillery, Mon.
G. N. Haden & Sons (Engineers) Caving Club:
7-12 Tavistock Square, London, WC 1

Hades Caving Club:
54 Ennadale Road, Langley, Nr. Macclesfield, Cheshire
Happy Wanderers Caving & Potholing Club:
Braida Garth Farm, Westhouse, via Carnforth, Lancs.
Harrow Mountaineers:
8 Lloyd Court, Westend Lane, Pinner, Middlesex
Harwell Exploration Group:
Icknield Way House, A.E.R.E., Harwell, Didcot, Berks.
Hereford Caving Club:
10 Glenmore, 293 Ross Road, Hereford
Hereford Caving Group:
18 Stretton Close, Fayre Oakes, Hereford
HMS Ariel Caving Club:
Mech. App. M. Williams, HMS *Ariel*, Lee-on-Solent, Hants.
Hollowford Training Centre Caving Club:
Castleton, Sheffield S30 2
Hope Valley College Caving Club:
Southwood, Lydgate, Eyam, via Sheffield
Hull Pothole Club:
17 Bristol Road, Wold Road, Hull, Yorks.
Hull Troglodytes Cave & Pothole Club:
Miss Carol Fenwick, 50 Chestnut Avenue, Queens Road, Hull
Hyperion Club:
12 Kelsall Close, Stockport, Cheshire
Hypogean Exploration Group:
R. Taylor, 143 Nottingham Road, Belper, Derbyshire

ICI (Welwyn) Rec. Club Caving Section:
ICI (Plastics), Bessemer Road, Welwyn Garden City, Herts.
Imperial College Caving Club:
c/o Imperial College Union, Prince Consort Road, London, W 7
Ireland, Speleological Society of:
c/o Department of Geography, Trinity College, Dublin
Irish Caving Club:
Miss F. Tinsdale, 21A Thomas Street, Portadown, Co. Armagh

Junior Soldiers' Wing Caving Section:
Dept. Wessex Brigade, Heathfield Camp, Hamilton, Devon

Kendal Caving Club:
Heathfield, 9 Sedbergh Road, Kendal, Westmorland
Kings Norton Caving Club:
220 Linden Road, Kings Norton, Birmingham 30
King's School Mountaineering Club:
King's School, Gloucester

Leck Underwood Exploration Society:

13 Sunnyside Avenue,
Tunstall, Burton-on-Trent

Leeds University Union Speleological Society:
The Union, Leeds 2

Leicester University Spelaeological Society:
Students Union, University of Leicester, Leicester

Leicester & Nottingham Caving Club:
John Atkinson, 8 Highbury Road, Leicester

Liverpool University Mountaineering Club:
Potholing Section, c/o Students Union, Bedford Street North, Liverpool 7

Llandaff Technical College Caving Club:
2 Heathwood Road,
Heath, Cardiff

Llangattog Caving Group:
2 Fitzadam Place, Adamtown, Cardiff

Llanishen Rover Crew:
4 Gleba Place, Llanishen, Cardiff

London University College Speleological Society:
Gordon Street, London, WC 1

Lord Weymouth's School for Boys Caving Group:
14 Church Street, Warminster, Wilts.

Macclesfield Caving Club:
D. Whyte, Nescar Street,
Macclesfield, Cheshire

Manchester University Dental Students Speleological Society:
188 Lloyd Street North,
Manchester 14

Manchester University Speleological Society:
The University Union,
8 Oxford Road, Manchester 13

Mendip Caving Group:
88 Seaton Gardens,
Ruislip Manor, Middlesex

Mendip Cave Registry:
Knockauns, Combwich,
Bridgwater, Somerset

Mendip Nature Research Committee:
9 Churchill Road, Wells, Somerset

Mercia Caving Club:
Mrs S. E. Bowden,
69 Chadsfield Road,
Rugeley, Staffs.

Mine Exploration Society:
12 Penwerris Terrace,
Falmouth, Cornwall

The Moles — 19th Harrow Rover Crew:
48 Kingshill Avenue, Kenton,
Harrow, Middlesex

Monmouth Exploration Group:
Rush Common House,
Abingdon, Berks.

Moorland Ramblers' Club:
28 Melville Road, Kearsley,
Bolton, Lancs.

Morecambe Rock Climbers & Potholing Club:
87 Westminster Road,
Morecambe, Lancs.

Moreley Cavern & Fell Club:
42 Oakwell Road, Drighlington,
Nr. Bradford, Yorks.

Mountain Club, The:
6 Trinity Rise, Stafford

Mullard Pothole Club:
5 Harlech Drive,
Oswaldtwistle, Lancs.

Nature Conservancy:
19 Belgrave Square,
London, SW 1

Newport YHA Caving Society:
2 Hove Avenue, Newport, Mon.

North Lonsdale Speleological Club:
44 St. Vincent Street,
Barrow-in-Furness, Lancs.

North Staffs. Speleological Society:
107 Algar Road, Trent Vale,
Stoke-on-Trent

North Wales Caving Association:
44 Brookside, Donnington,
Wellington, Shropshire

Northern Cave Club:
4 Hawthorn Avenue, Thornbury,

Bradford 3, Yorks.

Northern Cavern & Mine Research Society:
33 Gladhowe Avenue,
Roundhay, Leeds 8

Northern Caving Clubs, Council of:
11 Prospect Avenue,
Lostock Hall, Preston, Lancs.

Northern Exploration Group:
390 Colne Road, Burnley, Lancs.

Northern Pennine Club:
Greenclose House, Clapham,
via Lancaster

Northern Speleological Group:
Sutcliffe House, Giggleswick,
Settle, Yorks.

Nottingham Rover Caving Group:
3 Humber Road, Beeston, Notts.

Nottingham University Caving Club:
J. A. Dunford, Union Building,
University Park, Nottingham

Notts. Four Ways Club:
46 Florence Road, Gedling,
Notts.

Old Westonians Speleological Society:
92 Devonshire Road,
Weston-super-Mare

O.M. Mines Research & Exploration Group:
Hawthorne Villas, 39 Monk
Street, Tutbury, Staffs.

Orpheus Caving Club:
41 Narrow Lane, Aylestone,
Leics.

Oxford University Cave Club:
T. R. Cecil, Exeter College,
Oxford

Paisley Troglodytes Club:
2352 Dumbarton Road,
Glasgow, W 4

Pathfinders' Mapping & Cave Exploration Society:
26 Angus Street, Troedyrhiw,
Merthyr Tydfil, Glam.

Peak District Mines Historical Society:
Mrs P. E. Lunn, 28 Kenbourne
Road, Sheffield, S7 1NL

Peakland Archaeological Society:
2 Avondale Road,
Stockport, Cheshire

Pegasus Club, Nottingham:
P. Watkinson, 11 Grouville Drive,
Woodthorpe, Nottingham

Pengelly (William) Cave Research Centre:
Butterbrock, Harford, Ivybridge,
Devon; or Mrs A. Longman,
4 Heath Rise, Brixham, Devon

Peveril Underground Survey Association:
45 Sprinkfield Road, Birkby,
Huddersfield, Yorks.

Phoenix & Old Mine Club:
106 Sydney Street,
Burton-on-Trent

Pioneer Club:
Seale Hayne Agricultural College,
Newton Abbot, Devon

Plymouth Caving Group:
12 Radnor Street, Plymouth,
Devon

Preston Caving & Climbing Group:
5 Empress Avenue, Fulwood,
Preston, Lancs.

RAF College Potholing Society:
RAF College, Cranwell,
Sleaford, Lincs.

Rolls-Royce Caving Club:
14 Evans Avenue,
Allestree, Derbyshire

Royal Forest of Dean Caving Club:
M. Sterry, 9 Oakhill Road,
Mitcheldean, Glos.

Royal Military Academy:
Sandhurst Mountaineering &
Exploration Society,
Mr S. N. I. Palmer,
Ypres Company, RMAS,
Camberley, Surrey

Rugby Scout Caving Club:
53 Manor Road, Rugby,
Warwicks.

Sale Caving Club:
27 Keswick Road, Broomwood
Estate, Timperley, Cheshire

Salford R.C.A.T. Pothole Club:

D. Brandon, Students' Union, Royal College of Advanced Technology, Salford

Severn Valley Caving Club:
17 Woodside Road, St Anne's Park, Bristol 4

Sheffield University Speleological Society:
c/o Students' Union, Sheffield

Shepton Mallet Caving Club:
158 Coronation Road, Southville, Bristol 3

Shrewsbury Caving Club:
7 Cornelia Crescent, Portland Nurseries, Shrewsbury

Shropshire Mining Club:
2 Princess Gardens, Newport, Shropshire

Sidcot School Speleological Society:
Sidcot School, Winscombe, Somerset

Skipton Caving Club:
1 Sharpav Avenue, Skipton-in-Craven, Yorks.

Somerset Constabulary Caving Section:
Frome Division, Frome, Somerset

Somerset Scout Caving Group:
16 Bramley Road, Street, Somerset

Southampton University Caving Club:
University of Southampton

South Cheshire Caving Club:
M. S. Lockett, 88 Bedford Street, Crewe

South Town Caving Club:
20 Clifton Park, Bristol 8

South Wales Caving Club:
1/10 Powell Street, Penwyllt, Penycae, Swansea

South West Essex Tech. College Caving Club:
Forest Road, Walthamstow, London, E 1

Southern Caving Clubs (Council of) Withey House:
Withey Close West, Bristol

Speleological Society:
15 Westgrove Lane, Greenwich, London, SE 10

Speleo Rhal Caving Club:
85 Lower Northam Road, Hedge End, Southampton

Stoke-on-Trent Pothole Club:
10 Beech Drive, Clough Hall, Kidsgrove, Stoke-on-Trent

Streatham Senior Scout Caving Group:
63 Harborough Road, Streatham, London, SW 16

Swallows Nest Senior Scouts Caving Group:
83 Station Road, Woodhouse, Sheffield

University College of Swansea Caving Group:
Singleton Park, Swansea

Swansea YHA Caving Section:
3 Montpelier Terrace, Swansea

Tamworth Caving Club:
J. F. Follett, 36 Croft Street, Tamworth, Staffs.

Termes Caving Club:
P. W. Reavill, 3 Cater Close, Gedling, Notts.

Twenty One Club:
British Cellophane Ltd, Bridgwater, Somerset

Wells Cathedral School Caving Club:
J. Cowling, 5 East Liberty, Wells, Somerset

Welsh College of Advanced Technology Caving Club:
Cae Morgan, Cardigan

Wessex Cave Club:
Chaumby, Wookey Hole Lane, Wookey Hole, Wells, Somerset

Westinghouse Apprentices Caving Section:
79 Oaklands, Chippenham, Wilts.

West Midlands Cave Exploration Group:
J. E. Fell, 345 Jockey Road, Sutton Coldfield

Westminster Speleological Group:
The Railway Arms, Station Road, Theale, Reading, Berks.

West Surrey Caving Club:
19 Cardinal Drive,
Walton-on-Thames

Whitchurch Speleological Society:
Hillymead, 1 Llwyn Omn,
Pantmawr, Cardiff

White Hall Open Country Pursuits Centre:
White Hall, Manchester Road,
Buxton, Derbyshire

William Olds Boys' Club Caving Club:
J. M. McPhilbin, Chiltern Way,
Bestwood Park Estate,
Nottingham

Wolverhampton Caving Group:
8 Malins Road, Fighting Cocks,
Wolverhampton

Worcester Training College Caving Club:
M. J. Calvert, City of Worcester
Training College, Henwick
Grove, Worcester

Wrexham Caving Club:
N. Schofield, Sorby Hall,
Sheffield 11

Wyreside Potholing Club:
39 Poulton Old Road,
Blackpool

YMCA Rover Scouts Caving Section:
34 Birrell Road, Sherwood Rise,
Nottingham

Yorkshire Ramblers' Club:
Brierley House, Okenshaw,
Bradford, Yorks.

Yorkshire Underground Research Team:
Duke Street Cottages,
Greenhow, Pateley Bridge, Yorks.

Bibliography

British Caving, Routledge & Kegan Paul

Baker, E. A., *Moors, Crags & Caves of Derbyshire*

Barrington, Nicholas, *The Caves of Mendip*, Dalesman Publishing Co.

Casteret, Norbert, *Ten Years under the Earth*, Penguin

Coleman, J. C., *The Caves of Ireland*, Anvil Books

Cons, David, *Cave Craft*, Harrap

Expedition '67 to the Gouffre Berger, Pegasus Club, Nottingham

Ford, Trevor, *The Caves of Derbyshire*, Dalesman

Gemmell, Arthur, & Meyers, J. O., *Underground Adventure*, Dalesman

Holland, Eric G., *Underground in Furness*, Dalesman

Jasinski, Marc, *Caves & Caving*, Paul Hamlyn

Jenkins, D. W., & Williams, Mason, *Caves in Wales & the Marches*, Dalesman.

Johnson, Peter, *The History of Mendip Caving*, David & Charles

Kenall & Wroot, *The Geology of Yorkshire*

Lovelock, James, *Life & Death Underground*, Bell & Co.

Mountain Rescue Committee, *Mountain & Cave Rescue*

Nelson, A., & K. D., *Dictionary of Applied Geology*, Newnes

Nylon Rope & Climbing Safety, British Ropes Ltd.

Pennington, Rooke, *Barrows & Bone Caves of Derbyshire*, Macmillan

Porteus, Crichton, *Caves & Caverns of Peakland*, Come-to-Derbyshire Association

Robinson, Donald & Greenbank, Anthony, *Caving & Potholing*, Constable

Thornber, Norman, *Pennine Underground*, Dalesman

Index